2131: The Edge of Hope

By

B.E. Smith

HEMINGWAY
PUBLISHERS

Dedication

To my beautiful wife and precious children.

CONTENTS

Dedication   ii

About the Author   vii

Prologue

Chapter 1: Back to the Drawing Board   1

Chapter 2: Planning   12

Chapter 3: An Unwanted Guest   19

Chapter 4: Starting University – Again: (27.07.1981)   25

Chapter 5: Playing it Cool   36

Chapter 6: The Director   40

Chapter 7: The Mechanics of a Killer   44

Chapter 8: Preparation for the Call to the Big Leagues   49

Chapter 9: A Fresh Start (07.09.1981)   57

Chapter 10: New Territory (11.09.1981)   65

Chapter11: How to Blow a Prof's Mind 101   69

Chapter 12: The Future is Calling: (08.03.2135)   82

Chapter 13: Time Flies, when you're Racing it   84

Chapter14: Something's Horribly Wrong   94

Chapter 15: Blast from the Future (21.11.1981)   101

Chapter 16: Doctor Hoffman's Awakening   105

Chapter 17: Doctor Gus and the Bullet Hole   107

Chapter 18: The Rescue of Doctor Hoffman   112

Chapter 19: Damage Control (08.03.2135 12:15 hrs.)   119

Chapter 20: Catching a Shadow (23.11.1981)   125

Chapter 21: The Unexpected Time Traveler   134

Chapter 22: Act Normal; Whatever That is   138

Chapter 23: Revising the Mission   147

Chapter 24: A Brief History of North America Part I   150

Chapter 25: The Invitation to the Telescope   158

Chapter 26: Make to the Queen!   166

Chapter 27: The Life of a Killer   175

Chapter 28: Little Franny   178

Chapter 29: The Forensic Report   183

Chapter 30: You've Got Some Explaining to Do   197

Chapter 31: Mission Damage Control   206

Chapter32: The Steward Observatory   216

Chapter 33: Back to the Drawing Board - Again   228

Chapter 34: Fran's Unintentional Ruse   238

Chapter 35: The Sabbatical (30.11.1981 10:38 hrs.)   247

Chapter 36: Hacking the 1980's   256

Chapter 37: Fran and Maryam Meet with the Director   267

Chapter 38: The Extraction (01.12.1981 06:59 hrs.)   279

Chapter 39: The Mechanics of Moving a Monster   291

Chapter40: Getting out of Dodge   294

Chapter 41: Hunting a Scientist   302

Chapter 42: Leonardo and Kevin's Flight to Hawaii   304

Chapter 43: I Fought the Law   314

Chapter 44: Leo Steps Up to the Plate   332

Chapter 45: Getting Out of the Wolves' Den   338

Chapter 46: 1980's Policing   359

Chapter 47: Setting up at the CFH Telescope   370

Chapter 48: Getting out of Dodge - Again   379

Chapter 49: Watching the Stars   395

Chapter 50: A Brief History of North America Part II   419

Chapter 51: Bruce on the Run (04.12.1981)   425

Chapter 52: Scramble   431

Chapter 53: Half-Truths   449

Chapter 54: 8-1-8-2   459

Chapter 55: The Wrath of Shadow   471

Chapter 56: Doctor Mitch Smith's Dilemma   495

Chapter 57: Making Due   510

Chapter 58: Transportation of People, Goods, and Energy in the
22nd Century   534

Chapter 59: Honoring a Promise   543

Chapter 60: The Moon, The Mine, and The Mother   556

Chapter 61: The Intruder   564

About the Author

B.E. Smith has been captivated by science, physics, and astronomy since he was a child. His parents provided endless books, teaching lessons, and even a telescope to foster this fascination.

B.E. Smith attended Western University in Ontario, Canada, earning a Bachelor of Science degree in Applied Mathematics and Physics. Following university, he answered a calling and began a career as a first responder. After years of experiencing traumatic events, he began using creative writing as an outlet to cope with trauma.

He was selected for and successfully completed the Boots on the Ground PTSD service dog training program and now works with his service dog, Pax.

B.E. Smith lives with his wife and three children in Northern Ontario, Canada.

Prologue

Scorpions, lizards, and cacti were the only desert witnesses to the strange, otherworldly light that appeared and disappeared just as quickly, enveloping the man who had just appeared there. The beautiful iridescent blue light dissipated as it illuminated the tall, slim man's youthful complexion, weathered red ballcap, dark-rimmed glasses, and khaki pants.

The man bounced his backpack higher onto his shoulders as he scanned the familiar landscape - his emerald green eyes were unwavering - he trained for this for a long time. His determination to complete what lay ahead was a fire in his belly that rivaled the power of the blue light that transported him to this place.

As he began walking, the desert guardedly returned to normal, and the crickets and other night insects terminated their silence after several seconds. The man would return to this desolate place, but not for ten years, six months, and five days. That is if he was successful this time.

Chapter 1: Back to the Drawing Board

As the traveler walked past the sign for Willcox in the morning light, he shielded his eyes from the blowing sand and dust with the peak of his worn ballcap; he knew exactly where to go as he passed familiar trees, small shops, and old automobiles. He nodded to the people as he passed by, as he knew was customary. The people, in turn, stared at him quizzically, eyeing him up and down to take in his odd appearance: his dirty, freckly complexion and red-stained running shoes.

He carried all his worldly possessions in a black, weathered packsack that he slung over his left shoulder, his skinny arms concealing his considerable physical strength. As he walked onto Fremont Street, a dead-end street, he walked up to a tiny white house at the end with a wooden sign in the front with faded red paint lettering that read "For Sale."

The house had been abandoned for some time and showed many signs of its age and lack of upkeep; the white shutters with their peeling paint were leaning precariously, having become loose in the desert winds over time. It was wholly exposed to Mother Nature's desert elements since it was on the edge of Willcox on the South side of the town.

Commercial properties were creeping into the residential area, and only a handful of houses remained as if they were the contents of a time capsule. The old stone path to the house was littered with windblown desert sand, and the roof was sagging badly on one side of its peak. Without hesitation, he walked to the house next door and knocked, removing his glasses and tucking his polyester shirt into his pants. He was tired from his long early morning trek and wanted to get to work. He knocked at the door and waited.

"Wadda, you want stranger?" a woman yelled as she peered through the tiny crack she had opened in the door. He could see her tight, dark curls bouncing in the curlers beside her pink-framed glasses as she studied him up and down.

He cleared his throat. "I see a house for sale next door; you must know who owns it. I want to buy it."

"You'd be a fool to buy it!" she yelled. "Come back later this afternoon and take a look. You'll see it's not for you." She sneered, her lips pursed, as she peered through the crack.

"I just got to town and need a place to stay. I have come into some money. I will pay you twelve hundred dollars right now to rent the house for a year," he retorted, knowing full well that that kind of money was going to be irresistible to her, "and then once I get a job, I will buy the house," he finished.

There was another long pause.

The door screeched open. "You got that kind of money with you?" She paused as she got a complete up-and-down look at the stranger at her door. It's my house. Where you from anyways?"

"Yes, ma'am, I do have the money with me. I recently lost my father and left New York on the train a month ago. I like this town and would like to enroll in the school and stay for a while," he lied.

She studied him even closer, her eyes squinting as she thought. "So, what if someone else comes to buy the house while you're rentin' it."

"Then I will leave," he said quickly, "And if that happens, you can keep the money, and I will go."

She guardedly stepped out onto the crooked porch, her eyes fixated on his. "You sure you want to live in Willcox? It's nothin' like New York – so I hear."

He stood his ground as she stepped closer as if she was poised to straighten his shirt collar.

He stood his ground. "No, ma'am, it's not like New York; I just want some peace to sort out what I want to do with my life. Get a fresh start, ma'am, get an education and a job."

"Okay, sonny, but don't say I didn't warn you." She shook her head and smiled in disbelief at what was transpiring in front of her. "My name is Miss Rosie. I might be your landlady now, but we won't forget our manners and drop the 'Miss,' now, will we?"

"No, ma'am, you have my word," he stammered, keeping his eyes on the ground as he stepped off the porch and walked towards his new accommodations.

"Didn't catch your name, Sonny," she yelled after him. "Let me get the key for you."

"Bruce. Bruce Hayden, ma'am. A pleasure to meet you." His voice trailed off as she disappeared into her house. Miss Rosie returned a moment later with a single key on an old ratty string and walked over to unlock the brown wooden door. She paused before she put the key in the lock.

"I need that money this morning, Mr. Bruce," she said sternly, looking over her glasses, which were perched precariously at the end of her ski jump-shaped nose.

After walking inside the house, Bruce looked around at the familiar surroundings.

The tiny kitchen, with daylight streaming through the old single-pane kitchen window and the sitting area beside it, was just as he remembered. He leaned backward and glanced into the bathroom off the kitchen. The musty smell of the house had a tangible uniqueness to it.

It brought back so many memories from another life here, so many memories.

"Always be careful not to mix up where and when

you are, and be especially careful not to mix up events on a different timeline that you may have been a part of; otherwise, it will be taxing on your mind over time." Bruce's psychologist, Sandy's nagging voice rang in his ears; her warning had been issued to him many, many times in training. Those events seemed dreamlike now, the ones from the future.

He focused back on the task at hand.

"Looks like home." He smiled at Miss Rosie.

"The bedroom is upstairs, and a set of sheets is in the top dresser drawer. I trust you can manage from here, Mr. Bruce," she said.

"Yes, ma'am, I can. Thank you for your time, and if you allow me a minute to put my bag upstairs, I'll come right back with your money." He said it more like a statement than a question, but he was dog-tired and wanted to keep their first interaction quick.

Bruce got upstairs to the only bedroom and dropped his black bag onto the old, creaky, single bed where he was out of sight of Miss Rosie, and he unzipped it.

From inside a small zippered side pocket, he pulled

out a bundle of money with a green elastic; he was nothing if not organized. Being tediously organized in his work was necessary, or he would risk failure.

He trotted down the creaky wooden stairs two at a time and handed the bundle of money to Miss Rosie. "That should be the full twelve hundred, ma'am. I won't be offended if you count it before you leave," he said with a smirk as she grabbed the money with greedy restraint.

She held the bundle tight to her chest, obviously dumbfounded by her luck.

She was distracted, trying not to look at the money. "I'm sure it's all there; besides, you'll see me shortly if it's not." She smiled at him, and her shoulders relaxed slightly – she trusted him now.

"There is a small basement, too. I can tell you some history about this house, it has some neat history. My older brother lived in this house; God rest his soul." She walked towards the basement door.

"Maybe some other time?" he interjected, now halfway up the stairs. He did not want her to get on one of her storytelling diatribes; he was dead tired and needed to rest before completing his day's work. "Well, good day then, Miss Rosie. Thanks for your time," he yelled as he got to his bedroom.

She left the house and walked across the sandy lot to her place, her money clutched tightly in her hands. When she was out of sight in her home, no doubt counting her lucky stars at her windfall, Bruce came down the stairs and quietly locked the front door.

After shaking out the old dusty sheets and putting them on his bed, he lay on it and glanced at his watch; it was 9:00. He got up and tucked his black bag into the drawer of the old dilapidated wooden dresser in the corner of the room, after removing a smaller black pouch from one of the compartments. He clutched it in his hand; his long fingers could circumnavigate the bag easily as he felt the shapes of the familiar tools inside.

He leaned in carefully to access the dresser lest he crank his head on the inconveniently sloped ceiling that matched the roof's pitch; he had done so quite painfully many times before. He held the smaller satchel for a moment and gently shook it, mentally inventorying its contents while shaking it.

He needed his handyman kit to modify the basement space into the high-tech working area where he would spend so much time.

He yawned. It was counterintuitive to nap with so many tasks still to be completed that day; however, he needed to rest briefly and recharge after his four-hour early morning walk in the desert.

Bruce's watch vibrated on his wrist, indicating it was time to get up.

He had slept deeply and felt refreshed; it was 10:30. Grabbing the small pouch with his workplace conversion tools, he ran down the two flights of stairs to the basement as he had done so thousands of times. He thought he could do it with closed eyes but did not want to test the theory.

He got to work; first, he went over to the corner of the basement, where a small space was under the stairs, and he slid aside the old dirty curtain that concealed the upstairs area. He immediately spotted what he was looking for: an old toaster, two fluorescent lighting tubes, some loose low-voltage copper wiring, a hammer, and a hand saw. The other items strewn under the stairs were for another task.

He placed the toaster, light tubes, wiring, hammer, and hand saw on the old rectangular table, and he mentally went over the materials he needed to go and get at the hardware store and trotted back up the stairs to the dresser in his bedroom.

He put the black pouch back into his bag, slung the bag over his left shoulder, and was down the stairs and out the front door as he re-visited his mental list for the store. He walked out to the roadway and put on his old red ballcap; the wind blew sweet scents of flowers to his nostrils as he walked out the path to the street.

Miss Rosie will be ensconced in her fresh windfall and watching her daytime soap operas, he thought as he walked by her front lawn and down the street. Clarke's hardware store was about a kilometer away, but the residential streets had beautiful, tall trees that branched out enough to provide some shade.

For most others, his walking pace was the equivalent of a slow jog; his long, skinny legs had been brutally trained in the months before his mission.

He had a spring in his step as he took in the bird

songs and the music of the wind blowing through the tree leaves; Bruce enjoyed the fresh open air whenever he could get it. It would take him quite a few hours to transform the basement space into a highly functional space for him to work. The 22nd-century technology Bruce possessed would make the run-down house a technological fortress.

The basement's transformation on Fremont Street marked the beginning of his years-long mission, and he did not have a moment to lose. The fate of the world depended on it.

Chapter 2: Planning

The morning sun trickled into Bruce's bedroom through the old, painted, wood-framed window, shining on Bruce's face. He awoke with a start and realized where he was as the late hours of the previous night replayed in his mind. The black watch with its band vibrating on his wrist was plain at first sight; however, it was far more advanced than it looked for 1981.

With his watch, Bruce could track people's movements, record audio and video, and keep a research log, which automatically backed up to his tablet in the black bag.

The tablet, which appeared to be a hardcover book of Shakespeare's plays, was a sophisticated 22nd-century computer integral to his mission. The laptop was more vital than the small black spherical transportation device hidden in his bag. If essential for the task to be successful, Bruce would maroon himself in this place and not return to the future.

The transportation device, a billiards ball-sized spherical device called the Hoffman Einstein Rosen Bridge time travel device, or HERB, was solid jet black until activated and could only be activated by fingerprint pressure on its surface and facial recognition technology. Two 19th-century physicists, Albert Einstein and Nathan Rosen, theorized that, given the proper amount of energy, a wormhole could significantly shorten the distance between two points and alter the traveler's location in time. Doctor Leonardo Hoffman, a 22-century physicist, successfully proved Einstein and Rosen's theory and made it a reality.

A top-secret device in the 22nd century, the time travel orb project was called the HERB project. When the HERB was activated, the orb glowed a dark translucent blue color and could transport Bruce back to the past or forward into the future. A limitation of the HERB was that after transporting Bruce to 1981, there was only enough energy to return him to the 22nd century without recharging.

There was much debate amongst the scientists and supervisory panel on whether to send a mobile charging station back with Bruce and allow more than one return trip in an emergency. Still, the charging technology was too cumbersome and required fusion technology. They could not risk leaving the bulky charging device hidden in the desert because it was too large and heavy for Bruce to carry. He would be trapped in 1981 without a fully functioning orb and would be long dead before his wife was born in Ohio in the year 2112.

Bruce inventoried his belongings and set his bag on the bed. He took a moment to soak in the morning sunshine and review the plan for the day. Usually, he could not spend time in the searing Sun for long due to his pale skin. As a child, his red-haired complexion had always made him susceptible to painful sunburns. Thanks to an injection of multipurpose nanobots into his bloodstream before the mission, the intense UV rays of the Sun would not damage his skin - the nanobots would also protect him from dehydration, disease, infection, or trauma.

He walked down the stairs with his bag slung over his slim shoulders, his large feet turned on an angle to navigate the old squeaky stairs. He felt invigorated, fresh, and ready to work.

The basement air smelled stale, and there was a

thick layer of dust from years of neglectful cleaning. The smell of Bruce's electrical work from the night before invaded the staleness of the basement. There was still just a hint of the unique scent of old fermented grapes in the stale air, to go with a purple stain in one corner of the room where someone spilled homemade porch crawler-type alcohol.

He wiped the dust and small electrical wiring debris from the table with his hand, wiped the dust on his pants, and dropped his bag on the small wooden table.

He turned on the new light source in the room, which now hung by two cables, and swung slightly above the table, creating moving shadows on the wall; the airborne dust, freshly disturbed, revealed itself and shone in the new and improved soft LED light. The lighting upgrade he had completed the night before would decrease his eye tiredness and increase his concentration time.

Part of his work on day two was to enlarge the table to get the most productivity out of the cramped basement space. He removed his Shakespeare's plays book from the bag and opened it, revealing the 22nd-century computing device. The holographic touch screen occupied one side of the interior, while the rest of the book's thickness was actual Shakespearean plays.

The computer stored the plans and data he needed to accomplish his goals over the next decade; not only did he have to create technologies and computer programs that were closely compatible with 20th-century mathematics, physics, and engineering, but he also had to convince high powered people to fund his project that would ultimately save the world. Bruce read the display on his tablet and synched it with his watch with a glance at his watch:

Day 2 - morning: complete work to transform basement space. His watch was wholly automated - it would keep track of hours worked, the efficiency of his work time, concentration levels, and calories expended. As Bruce began working, the world outside his little basement in the old house at the end of Fremont Street was oblivious to the otherworldly events that were taking place.

Hours swiftly passed as Bruce worked, and he was surprised when his watch vibrated, instantly interrupting his concentration and alerting him that he needed to go and take a walk for a break.

The afternoon sun felt nice on his face as he soaked it in while contemplating the countless hours he would spend in his basement, creating new technologies necessary for his mission. He would have to put on a terrific act to convince the top minds of the scientific community that he had made the new technologies himself and not from the future. One hundred fifty years in the future, to be exact.

Bruce Hayden was selected for the HERB mission from a large pool of candidates, mainly young scientists and military personnel. He underwent rigorous psychological, technological, intellectual, and physical testing. As the candidates were whittled down to a handful of young men and women, one of Bruce's friends, Lee, a competitor for the mission, joked during training that Bruce could probably make a small nuclear fission reactor from an old toaster oven. He would have to pull off a similarly lofty feat to accomplish his mission this time; the task was highly complex and fluid and would span more than a decade. He learned a multitude of lessons in the previous two times he attempted to save Earth, but he failed by two months the first time and failed by sixteen days the subsequent time.

The board of directors made it painfully clear to

Bruce that this was his last attempt. To secure his spot in the third mission attempt, he had to show the selection committee that he had the schematics, plans, and schedule to memory, including the timeline for making scientific discoveries at the university.

What ultimately separated him from the other three final candidates was that he had lived the mission two previous times and knew all the major players.

All the scientific discoveries that Bruce would accelerate could not all come from him; otherwise, it would become unbelievable. In a nutshell, he had to progress the day's technologies as fast as possible, with the least amount of pomp and circumstance. Trained to take things one day at a time and compartmentalize his mission instead of looking at the entire multi-year plan - if he were to have a nervous breakdown or become seriously injured at any point - he would surely fail. Humanity was running out of options, and Bruce's failure would likely mean the extinction of the human race.

Chapter 3: An Unwanted Guest

Asteroid Smith-Gregson, also known as SG-2131, moved silently and invisibly for countless eons, unnoticed; an invisible planet killer traveling at speeds up to 16 kilometers per second in an obliquely elliptical orbit around the Sun.

Composed of mostly iron and carbon, it was almost 13 kilometers long and 8 kilometers at its widest: in its oblong football shape, it tumbled through space end-over-end with a slow, 22-hour-long rotation about its long axis. Its dense composition and sheer size made it the killer that it was; a comparable asteroid wiped out the dinosaurs; however, SG-2131 had a significantly higher density, which meant it would slam into the Earth with exponentially more energy.

An international group of scientists that had evolved from the European Council for Nuclear Research (CERN), operating on the Franco-Swiss border in the late 21st century, were using an underground facility remotely attached to CERN, known as the Highly Experimental Physics Area (HEPA). Known only to the handful of scientists working at HEPA, they developed a prototype for time travel technology late in 2098.

It was a crude prototype, and experiments over the years were methodically expanded from a few seconds of travel up to an hour. It was nothing like the time travel device that Bruce had used to travel back 150 years - that was the product of another thirty-three years of intensive research and experimentation.

At its peak in 2131, the funding required for the HERB project became so immense that it had to be financially backed by several key countries, even though they didn't know the nature of the research nor the HERB mission. Creating new technologies that were spinoffs of the HERB project and selling patents for the latest technologies were significant sources of its income and provided ample incentive for the countries financially backing the HERB project.

By the time SG-2131 was discovered in April 2131, and it was determined by computer simulation that Earth would sustain a direct hit by the monster seven years later, HEPA began formulating a mission to divert SG-2131 by sending a person into the past to discover the asteroid when it passed Earth in its previous orbit, thereby diverting it from its trajectory.

Scientific groups initiated parallel projects to divert the asteroid from its path. NASA launched all the landing pods it could create as the time crunch to move the monster became increasingly desperate. The landing pods were used before in experimental situations, and the science of moving large orbital objects was well known; however, the nearly three dozen fusion plasma jets all firing at maximum power for the seven years remaining was not enough force to move the asteroid.

NASA touted to the public that they could divert the asteroid enough that it would pass through the atmosphere and back out into space or have a merciful glancing blow that may not cause as much damage to the Earth as a direct hit.

It didn't take long for the rest of the scientific community to prove using computer simulations that even with their advanced technology to launch and land the pods rapidly, even seven years of thirty-six active fusion plasma jets was not nearly enough to change the orbit. Most simulations confirmed that the time needed to do so required at least double that time.

With extreme pressure from the government, most scientists quickly agreed that they would continue to work on the problem without illustrating the minuscule chance that humanity would survive the asteroid strike. News stations dedicated hours and hours per day to exclusive after exclusive programs related to the creating, launching, landing, and firing of the fusion plasma jets - sadly, even when humanity was staring down the barrel of certain death - people were trying to capitalize and make money from it.

Classified as a planet killer, SG-2131 would slam into the Earth at a velocity of 12 kilometers per second, having dumped a negligible amount of its kinetic energy in the Earth's atmosphere. It would hit the Earth with such force that it would punch a hole through the Earth's crust and spew lava, rock, water vapor, and dust around the globe. Some scientists argued that if the killer struck on the long end like a football spiral, it could pierce the Earth deep enough to spew the contents of the Earth's molten liquid outer core onto the surface, killing all animal and plant life.

The soot and ash would rocket into the upper atmosphere and block the Sun for hundreds of years. The steam from the molten rock evaporating the oceans would propel debris and dust into the stratosphere. Mega quakes would rock the Earth for several months or years as the wounded Earth shifted and bled molten rock. SG-2131 would cause the extinction of sentient life on the planet; almost all independent computer simulations confirmed this outcome.

Microbial life would survive, even thrive. However, the persistence of homo-sapiens would not be left to a slight chance. Other programs were modified or accelerated to collect DNA from the world's population of humans, plants, animals, and insects. These other plans were to deposit the cryogenically frozen DNA in a storage facility on the Moon in the hopes that the small civilization living in the moon colony could re-establish life on Earth after it became habitable again if it became habitable again. The media was in an absolute frenzy about these plans to save humanity, which provided a great distraction while the HEPA team secretly finalized their plans.

Passing by the planet Jupiter, SG-2131 silently accelerated towards the Sun and passed between the Earth and the Moon in 1989. Passing at a distance of 100,000 kilometers from the Earth, it was an astrological close shave. However, the people of the 1980s were oblivious to its existence, and the massive SG-2131 would pass by the Earth unnoticed, traveling back out to the solar system's outer reaches before returning and setting its sights on Earth for a direct blow in the year 2138.

For now, the killer drifted along inconspicuously, camouflaged perfectly against the black night sky. On his mission, Bruce Hayden would be a student working as an astronomer's assistant at an observatory in Arizona, where he would discover SG-2131, called Nault-Hayden-1983 or NH-1983 in the history books, should Bruce be successful.

Nobody else in the world knew it, but they were all counting on him.

Chapter 4: Starting University – Again: (27.07.1981)

The University of Arizona campus was a perfect place for Bruce to start his university career in 1981; it had a large telescope facility in a remote location, capable of making an asteroid discovery.

The university campus in Willcox was a small satellite-type extension of the main campus in Tucson. This smaller campus consisted of a few large buildings with a correspondingly small number of faculty professors.

It was a lesser-known university campus for Astronomy and Physics in 1981, making it a desirable place for Bruce to work. It allowed him to enroll with no high school transcript, just a great cover story. He had fooled the faculty before into believing that he was a math and physics whiz who moved from Canada after his parents were killed in a car accident.

The mission panel had discussed forging a transcript for Bruce to use, and Bruce could hack in and alter the high school's software system; however, a phone call from the university to the high school could prove an unnecessary risk should none of the High School teachers, who's courses Bruce took, remember him.

Ultimately, Bruce applied as planned with no high school transcript. His application process had become drawn out in his previous two lives here; Bruce had to complete a series of tests and then work at an apprentice job before they would accept him into the Physics and Astronomy Undergraduate program. He could not waste that amount of time again.

Bruce walked up to the admissions desk, his application form in hand, and spoke softly but confidently.

"I'd like to enroll in the Physics and Astronomy department Undergraduate program in September."

The middle-aged lady behind the desk, who hadn't looked at him yet, held out her hand and said robotically, "Admission forms and high school transcripts."

Bruce held his breath for a moment. He could make this go smoother this time and save himself a little time. "Uh, here is my application…maybe I could see somebody from the Astronomy or Physics department to talk about my high school studies?"

The lady looked up at him, lowered her hand, and scanned the application he had handed her. "You don't have a high school transcript... Bruce?" she asked as she looked at the name at the top of his application.

"I went to school in Canada, and……," Bruce started.

"Pretty sure they have high school, too." She looked at him with one raised eyebrow. She was an attractive middle-aged woman with long brown hair and piercing blue eyes. She had been working at the university for a long time and recently moved to the smaller Willcox campus.

"If I could just speak to a physics and astronomy Professor, please," he begged.

"Sorry, you can't apply without a transcript. I don't make the rules, Bruce," she said, her blue eyes softening.

Time to turn things up; I've got nothing to lose. "Well," he stopped, his eyes welling up, "both my parents were killed in a car crash years ago, and I was home-schooled by my uncle, but I assure you that I am an excellent student – aren't there any students here who have been home-schooled?" He was rambling, getting louder, and crying visibly.

She put up her hand. "Oh, please," she begged. "Let me talk to my supervisor about home-schooled students, but it's very unusual."

Wow, he thought as he wiped one of the tears rolling uncomfortably past his nose. *Maybe I missed my calling in theatre.*

He could see the lady talking to a man in the back area of the office behind a large window. As she pointed to Bruce, he gave a sad wave and wiped the crocodile tears from his eyes. The man did not look impressed. He walked over with the lady trailing closely behind.

"So, you were home-schooled, Mr....." The man walked out of the office and into the hallway with Bruce.

"Hayden, Bruce Hayden, sir." He jumped into step with the man as they walked up the deserted hallway.

The man took a long look at Bruce and stopped. "In Canada?"

"Yes, sir, my uncle continued my home school

studies in Canada after my parents were killed in a car accident ten years ago." The man stared at Bruce for a long time. There was still a tear in Bruce's eye, and he let it glisten there for the man to see.

"All I want, sir, is to speak to a member of your Physics and Astronomy department. Is Doctor Nault available? I have read a lot about your school, sir, and I promise you, doctor Nault will not be disappointed if I can have five minutes of his time."

"How do you..." the man looked perplexed, staring at Bruce curiously.

A tall, skinny man with an English accent came up from the back of the office to the counter. "Did I hear my name?"

Bruce was amazed at Doctor Nault's appearance, seemingly from nowhere. This had never happened previously. Bruce was excited at his good luck.

"This young man was asking to see you, doctor Nault. He wants to enroll here in September, but he doesn't have a high school transcript of any kind."

"Are you Doctor Nault?" Bruce asked excitedly,

"Doctor Kevin Nault?"

The professor nodded, "Yes, my boy, I am – do I know you?"

"No sir," Bruce continued, getting down to business. "I want to talk to you about joining the Physics and Astronomy department at your school, and if I can just have five minutes of your time, you won't be disappointed." Bruce was sweating now. These events could be a big break for him and could save him weeks, if not months.

"The physics and astronomy program at this school at the undergraduate level is challenging, and unless you have a solid base in physics, you would not be successful in your first year, let alone in the program," Doctor Nault said sternly. He always had a knack for telling people how it was, and he did not mix words.

"I assure you, doctor Nault, I will be successful. If you give me five minutes of your time." Bruce begged. *This could go either way*, he thought. Kevin Nault was a famously busy man who spent all his available time researching when he wasn't teaching, but he was also curious and open-minded.

The supervisor was becoming impatient with Bruce. "Doctor Nault is a busy man; he doesn't have time for your games."

"This is not a game, sir," Bruce replied sternly. He meant that with every fiber of his being.

"Come with me, son." doctor Nault smirked as he started to walk farther down the hallway.

Bruce could tell that Kevin was intrigued. As they walked back to the Physics department, Bruce planned out what he would say to get Doctor Nault to accept his application.

It was something an 18-year-old wouldn't usually be proficient at, but not too advanced for 1981.

They walked into Doctor Nault's office, and Kevin sat behind his desk; his chair moved backward slightly on its wheels as he leaned back with his hands up and his index fingers pressed on his upper lip, thinking. The young doctor would have completed his post-doctorate studies at the University of Cambridge just two years ago and was heading into a significant era of research. "So," he put his hands down on his lap, "tell me what you know about the theory of quantum computing, Mr. Hayden." He had been looking at the cover of a physics magazine on his desk as he thought.

Jackpot! Bruce hid his excitement with a stoic look and composed himself to speak.

"Well, sir, I have recently read about doctor Richard Feynman, who has theorized quantum computing as an extension of quantum mechanical principles. I believe that it could revolutionize computers and the way they solve complex mathematical models. Quantum computing, as I understand it, could increase the speed of cumbersome calculations exponentially; calculations that would previously not be practical or impossible to compute."

He trailed off as he realized he was looking out the window and not at Doctor Nault, whose mouth was open and was staring at him in disbelief. There was a long pause. Their eyes locked when Bruce's gaze returned to the doctor's office.

Too much? Bruce wondered if that was over the top, too much information. He waited and looked at Doctor

Nault uncomfortably. He could feel himself blush. His genetics came through every time, the crimson red cheeks of a red-headed young man.

"Mr. Hayden, how…how, exactly do you think quantum computing would work?" Doctor Nault looked frantically around his desk, dumbfounded, and was probably looking for a sip of water or coffee to take in what he had just heard the young man say, or more likely, a paper and pencil to write down what he had just heard.

Bruce knew he could not go one inch further into detail, or he would be telling Doctor Nault details of a computing technology that had not yet been invented.

"Well, uh, sir, that is my belief and not facts what I just said, but I would dearly love to be able to study here in the Physics and Astronomy Department at this fine institution so that one day we could know the exact answer to your question."

Doctor Nault was a genius in his own right and was an excellent judge of character. He sat and stared longer at Bruce; he was thinking. Bruce suspected the doctor wasn't considering whether he would be admitted to the program but the implications of the hypothesis that Bruce had just introduced to him.

The question was, would Doctor Nault let Bruce into the Physics and Astronomy program starting in September, just weeks away, or would he have to work his way up to the program at other jobs at the university and waste time? *Again.*

The doctor broke his silence with a laugh, "Bruce, I love how your mind works and how you hypothesized an application of a theory that a Nobel Laureate just put forward. He held up the magazine *Physics and Astronomy, July 1981*, as he smiled. I will insist that you take an entrance exam this afternoon so we can be sure of what I know already."

"I would be happy to take an exam today," Bruce stammered, "but what do you already know?"

Doctor Nault smiled, "I have a feeling that you will do great things in your lifetime, and I want this university to be a part of those great things."

You don't know just how great those things will be. Bruce thought as he smiled back, "So where's this exam?"

"Ha! Right to business, kid, I love it!" Doctor Nault got up and rifled through some papers in a manila folder in the top drawer of a nearby filing cabinet.

"Here you go," he said as the one-page exam glided onto Bruce's lap. "Use that table over there in the corner. There's a pencil and eraser over there, too; you have an hour." Bruce looked at the paper. He worked hard to hide his excitement; this would be the best start yet. He smiled as he started to write some familiar formulae and concepts on the paper.

Chapter 5: Playing it Cool

Bruce left the university at a quick pace, his lanky legs taking long strides.

Once he was out of sight of the school, he broke into a run and let out a 'whoop' as he jumped and then glided down the streets towards his house; he honestly couldn't think of a better start to his mission, and he was ecstatic.

He arrived at the house and gave Miss Rose a wave as he passed by, got into the house, and locked the door behind him; the advanced security system automatically re-engaged as he entered his house. He went downstairs, sat at his corner desk, and turned on his mission log device to record the day's events.

Once completed, Bruce felt an exhilaration he hadn't felt in some time, and he checked on the nano camera he had conveniently left in Doctor Nault's office.

He had attached one of the 22nd-century devices to the side of Doctor Nault's Physics doctoral diploma on the wall. That diploma would not move from that wall until Kevin retired.

The nano camera Bruce had installed was something from science fiction to a person in 1981. The camera was as small as a speck of pepper and could easily transmit every sound, conversation, and movement in multiple wavelength bands for at least ten years.

As Bruce looked on, he could see Doctor Nault walk into his office carrying one piece of paper; he sat in his chair, examining what Bruce had written on his exam paper less than an hour ago. The exam hadn't been a test of regurgitating formulae as many exams tended to be; this was an exam that tested the ability of the student to improvise and think outside the box. Coming from another century, Bruce was the definition of a student who could think outside the box.

Bruce's gaze left the screen where Doctor Nault sat reading. He reviewed his daily log summary and Daily Integration Required Tasks, or DIRTS, as he had joked while in training, an acronym that described them well. He considered DIRTS sweeping up the day's dirt and compiling it as a video log. AI embedded the vital log information in multiple algorithms and simulations to guide him on his mission. He smiled as he stared at the video as it replayed on his tablet:

"I wonder if Kevin will let me start apprenticing in his lab starting the first year? He has always insisted on not allowing any undergraduate to enter his lab as an assistant without a solid two years at the school. Maybe 1, but a brand-new first-year student?" His voice trailed off and then disappeared as the logged video stopped.

Bruce rubbed his sparse, wiry red chin stubble with his thumb and fingers, staring off into the distance. The possibilities were so great with the fortuitous start he had this time.

"I'm going for it!" He exclaimed as he pounded his fist onto the desk, slightly shaking the holographic unit that sat there, inactive. He stood up and paced back and forth in the basement area, moving his hands and fingers as he worked out his plan.

It was always highlighted in training that moving too fast with the mission, which was a natural thing to want to do, was dangerous. All the data analysis and petabytes of computer simulation modeling compiled over the years of preparation and decades of previous mission data conclusively showed that this was their best chance, in the best location, with the best candidate.

Now, Bruce was perched to clinch the best start to the mission yet. He rubbed his hands over the top of his head from his face, took a deep breath, and exhaled it forcefully. Bruce visualized the pressure of the mission dissipating through his breath.

Don't let the mind engage the thought of how important this is, he thought as he exhaled through his pursed lips again.

Outside, nature's orchestra played; the birds sang their evening salutations harmoniously with the insects, and the desert breeze ruffled the leaves on the trees. All were oblivious to the stranger inside the decrepit house planning to save the world.

Chapter 6: The Director

The director of the HERB mission and the HEPA lab was an introvert in every way. Rising to become the director of such an influential panel of people came at a considerable cost; he had jockeyed, manipulated, and cajoled his way to the top of the supervisory board for the most important mission in human history, but in the process, he had lost his wife and alienated his two adult children.

Now, alone most of the time, the director only surfaced from his underground office space to attend meetings. His underground office space was more than just an office; it was his home.

The 2500 sq ft space included sleeping quarters, a kitchen, and an exercise area connected to his office space. His ultra-secret security clearance gave him great power, and his Swiss citizenship augmented that power, as he could make deals with foreign governments virtually unchecked.

He was accountable to no one except for giving

scripted updates to governments who had invested unconscionable amounts of money into the HERB program. Although none of the governments knew the exact nature of the mission, particularly the existence and usage of time travel, they knew that the HERB program would save the human race from Asteroid SG-2131 by any means necessary. The director's usual pitch to secure ongoing funding for the program was to say that the HERB program was developing technology to divert or destroy the Asteroid. It wasn't exactly a lie. Half-truths and secrecy were how the director had to operate; his true identity was never disclosed to any government or persons outside the HEPA lab.

The supervisory panel for the HERB mission, including five men and five women, and Doctor Leonardo Hoffman, the lead scientist in the lab, were the only people on Earth who knew the director's identity. The HEPA mission operated in complete secrecy; espionage, terrorism, and other extremist groups were a constant threat to the mission's success.

The director's brown eyes, pock-marked face, and bushy eyebrows accompanied his long black goatee and shaved head. He bore the marks of his past with pride and confidence: the thick, long scar on his scalp and jagged cuts down the left side of his face and neck. He had come to the HERB supervisory panel after his decorated military career was ended abruptly when an extremist on home soil attacked him. The attack left him a physically and emotionally broken man, but he was a survivor in every sense of the word. He had spent a year in a series of hospitals and had overcome a partial paralyzation and the subsequent emotional trauma after the attack. In addition to the scarring on his scalp, face, and neck, he walked with a slight limp; however, he was still a strong man.

His work on drone technology and nano weapons during his military career was the scientific component of his qualifications for leading the HERB mission. Always a scientist at heart, there was no part of the mission that he didn't understand, except for the complex mechanics of time travel; that was the job of his scientific team in the HEPA lab. His understanding of all facets of the mission and his cold and calculating decision-making made him a good director. He didn't visibly step into the different processes going on around him unless it was necessary. Still, he had all the relevant information streamed to his office all day and night to keep him updated on everything. Having his finger on the pulse of all aspects of the HERB mission was the director's life.

Chapter 7: The Mechanics of a Killer

Many asteroids orbiting the solar system originate in the Kuiper belt, which is a cloud of debris that extends from the orbit of Neptune to approximately 50 AU (Astronomical Units, 1 AU = distance from Earth to the Sun) away from the Sun. Asteroids have been bumping and jostling, mingling, and coagulating since the birth of the solar system; asteroids are acted upon by gravitational forces such as passing stars and the Milky Way galaxy itself. Occasionally, one of these large objects is pulled by invisible forces, which causes it to leave the Kuiper belt and begin an orbit into the inner solar system.

Since gravity is inversely proportional to the distance squared from the object, once the asteroid completes its path, accelerates towards the Sun, and speeds around it, it starts its journey back to the outer solar system. It is gradually slowed and pulled back again, which brings it again and again in its orbit into the inner solar system where Earth's orbit is.

SG-2131, classified as an M-type asteroid due to its metallic and rocky composition, was most likely shouldered by another massive asteroid and pulled ever so slightly by gravity to begin its fall toward our Sun.

Its dense nickel-iron composition resulted from the collision of two large objects: a protoplanet and an asteroid. The chance collision between the two, millions of years earlier, saw the protoplanet hit so hard by the asteroid that the protoplanet's dense core became detached from most of the rest of the planet.

Over the next million years, the protoplanet's heart was struck, and struck, and struck, in the billiards room of the cosmos, knocking away all of the more fragile rocky parts, resulting in the heavy metallic asteroid SG-2131 that was discovered by humans millions of years after that.

The gravitational pull that acted on the massive asteroid was ever so slight initially; however, as it began getting closer and closer to the Sun, its velocity increased to sixteen kilometers per second when it flew through the inner solar system like a wrecking ball. The mechanics of changing the orbit of such an asteroid was well known in theory and had been tested by the 21st century. Time was the most significant impediment since the orbital period of asteroids from the Kuiper belt was up to 200 years.

The conundrum of changing the orbit of such a fast-moving, heavy object was twofold: landing a spacecraft on the asteroid required the craft to have the same speed and direction as the asteroid, and once the ship had landed on the asteroid, a steady force on the asteroid over decades was required to move the orbit significantly.

In the early 21st century, it took over two years for a craft launched from Earth to land on an asteroid, requiring several engine firings to pinpoint the correct travel vector. The engineers of the time lacked the guidance systems and jet propulsion technology for the craft to land safely and deploy on the asteroid promptly.

A rocket fired from Earth's surface travels approximately 11 kilometers per second to break the Earth's gravitational pull. So, with an update in their tech courtesy of the 22nd century, Bruce could land and deploy a series of crafts on SG-2131 in the optimum position to push it off its natural course.

The problem Bruce faced was not the technology of synchronizing the 16 kilometer-per-second vector of the landing craft to land and secure itself to the asteroid; it was the technology required to exert a constant force on the asteroid over a long period. Finding fuel for these jets using 1980s technology was impossible due to the sheer energy they would require to fire them for so long.

Under normal circumstances, a regular 12-foot-long rod of uranium pellets lasted about six years in a typical nuclear reactor; Bruce would have to improve the efficiency significantly and miniaturize the nuclear reactor jets to move the monster. Six years of force from regular nuclear-powered jets pushing on the asteroid was not enough; Bruce needed more than six times that amount of reactor energy.

The barely significant nudge by each pod on the mammoth asteroid spanned over decades would move the orbit just enough to cause the asteroid to miss Earth and prevent its destruction, and Bruce had one shot to hit his mark in the year 1989 to save the Earth in 2138, 149 years later.

B.E. Smith

Chapter 8: Preparation for the Call to the Big Leagues (30.07.1981)

In the weeks after being accepted into the Arizona State University program, which started in September, courtesy of doctor Nault's entrance exam and recommendation, Bruce was organizing, planning, and cross-referencing past missions; he had come up with a slightly accelerated plan of action for beginning school in the Physics and Astronomy program.

AI had integrated what had occurred thus far and then extrapolated a timeline for Bruce's progress at the university using advanced simulation programs. These simulations had many petabytes of information and data from Bruce's previous missions, including Doctor Nault's behaviors, the physics and astronomy research, and the computer technology used and created in the 1980s.

Bruce poured over the holographic data looming over him in his basement space, the data progressing across the dark room with a flick of his finger to indicate to the computer to go to the next page. He stared for some time at the three-dimensional animated depiction of the orbit of SG-2131 as he rapidly flipped the holographic pages underneath and skimmed pages of plans and proposals, schematics, and computer systems code.

Computational simulations were in their infancy in the 1980s due to the slow, bulky computer processors at the time. A supercomputer in the 1980s filled many large rooms with processors using cumbersome, brute force technology; comparatively, a cell phone by about the year 2020 had far superior computing power.

After 2020, the efficiency of computer technology increased exponentially for another century before Bruce's life. Bruce's watch, coupled with his tablet, compared to a 1980s supercomputer, was like racing a rocket against a hamster.

The problem with such a disparity between the technologies was that Bruce needed to strap that hamster to the rocket to develop and test computing technologies that had yet to be invented. Bruce could control the missile for the hamster, but that hamster had better hold on for dear life.

Similarly, in 1981, the knowledge of Near-Earth Asteroids was a result of dumb luck rather than technology:

The night sky is riddled with comet debris much of the year, but most meteors are fragments of comets resulting from asteroid collisions within the solar system. The known asteroids that cross the orbital path of Earth are known as Earth-crossers and number over 10,000, but the asteroids that fall into the heavyweight category, like SG-2131, that are larger than 1 km in length or width, number less than 1000.

Observing near-Earth asteroids in the 1980s was practically non-existent. Project Spaceguard, a name taken from a 1970s Arthur C. Clarke novel, would not begin searching for near-Earth objects (NEOs) until after 1992, more than a decade away. Spaceguard was one of the programs that Bruce and Doctor Nault were going to have to pioneer.

Congress in the United States was not on track to pass the appropriate legislation to start Project Spaceguard until 1992. Bruce and Doctor Nault would accelerate the implementation of Project Spaceguard. Observing and cataloging NEOs required multiple observatories recording data of an object over a long period to determine its orbit, and the technology to see and track them in 1981 left much to be desired.

Fortunately, the observatory used by the University of Arizona was capable, with its existing setup, of observing SG-2131 when it started to become a speedy, blurry black dot in July of 1983; it could only be seen when it crossed and interrupted background starlight. One major thing that Bruce had to his advantage was the knowledge of SG-2131 and, most importantly, where and when to see it. He would have to make it look like a coincidence that he had observed it, and he would work in tandem with other observatories over subsequent weeks to show the calculation that the asteroid would strike Earth in 2138.

For Bruce, locating and cataloging NH-1983 would be the easy part. Convincing the world that the asteroid would strike Earth in the year 2138 would be significantly more complex; at first, it would be a probability argument, complete with a public backlash about using taxpayers' money and multiple conspiracy theories regarding the project (none of them picked 'man from the future traveled back in time to divert asteroid,' but they came up with some good ones) but, with the public panic that would ensue with Bruce and doctor Nault's calculations, and resulting presentations to Congress illustrating the risk the asteroid posed to Earth, Congress would have no choice but to start funding Project Safeguard.

Bruce disliked that part of the mission the previous two times; he had been stalked in Washington by a woman who was convinced Bruce was her long-lost husband, and Doctor Nault was the subject of an assassination attempt. Even by 1983 standards, public attention was high. Bruce did not like the limelight, and the limelight was not good for his mission. The advantage of public awareness in 1983 was that it was limited to news camera crews and groups distributing flyers and printing news articles.

Unfortunately, with such a media frenzy, multiple churches and extremist groups who had significant followers interpreted the asteroid as a sign from God, consequently causing tragic mass suicides.

1981 was chosen for its proximity to the closest fly-by of the asteroid, and conveniently, for the technology already invented at the time, and for the opportune lack of social media platforms.

Social media, when created in the early 2000s, was at first a great way to connect families and communities; however, they ultimately led to the misinformation wars that launched countries like the United States of America into civil wars that lasted well into the 2030s.

The year 1981 was the only time available to HEPA for the mission to be successful since the next previous pass of SG-2131 was in 1840 when rockets and computers were not invented.

In the months and years after the discovery of the asteroid, Bruce and Kevin would have to address the United States Congress more than once regarding the asteroid diversion plan. At the same time, Bruce would keep the utmost secrecy regarding the exact details of his work. He would gladly pass kudos over to the Space Telescope Science Institute at Johns Hopkins University in Maryland, as well as Canadian, English, German, and Australian astronomers to keep the hyped and panicked eyes of the world off him and Doctor Nault.

The public could panic all they wanted, he thought at the time. *Just stay the hell away from us.*

The sudden onset of panic and the public attention to the start of the Safeguard program would slowly dissipate over the months and years since all the people living in 1983 quickly came to terms that they, and likely their children, would all be dead by the time the asteroid was to strike the Earth in 155 years. "It was a project to save the human race," he would say to Congress, playing to the American ego of the time.

Multiple countries and some key billionaires would fund Project Safeguard, paving the way to start the asteroid diversion part of the mission and create landing pods to divert the asteroid. Keeping the Safeguard name had been discussed multiple times in preparation for Bruce's tasks in the spirit of changing the past to the most minor extent possible.

In the first two years of Bruce's Undergraduate degree program, he would have to show Doctor Nault how he would develop a computing algorithm and software to observe, track, and incorporate other observatory data to calculate a NEO's precise orbit. Doctor Nault was an essential asset to this mission, even though he was unaware of his role in the most important scientific project in humankind's history.

Chapter 9: A Fresh Start (07.09.1981)

"Good morning, everyone," Doctor Nault sang as he waltzed into the classroom, addressing the new first-year students in the Physics and Astronomy program. He loved teaching first-year students since they had fresh, open minds, ready to learn. Doctor Nault's eyes scanned the classroom that Monday morning as he smiled. Many of the students smiled back; however, doctor Nault was looking for a particular individual.

"Ah," he said with a smile and a head tilt as he set his eyes on Bruce, who was in the front row, book open, pencil at the ready. Bruce gave a little nod and a smile back to the professor.

"Welcome all of you to the start of the year and the start of your university career." doctor Nault smiled at the class as he raised his chalkboard brush in a toast. Bruce could tell his beaming mood was not just because he was excited to start with this first-year Calculus class but at the prospect of what lay ahead.

You have no idea, Kev. You'd better strap yourself in, Bruce thought as he flipped through the first few pages of the textbook.

The class progressed, with Bruce writing notes completely on autopilot; the concepts and formulae of rudimentary calculus had been conquered by Bruce by the time he was eight years old. A bright boy recognized at a young age by the academic community, he was given enrichment lessons in physics and calculus, as much as he could handle, which was a lot.

By twelve years of age, Bruce had completed his double undergraduate degree in experimental physics and Engineering from Western University in Ontario, Canada. By fourteen, he began working in experimental high-energy particle physics and fusion technology with Doctor Leonardo Hoffman, a brilliant professor who introduced him to the scientific minds at HEPA.

Some of the scientists felt that fourteen years of age was too young to be privy to the ultra-top-secret technology that was HEPA, but Bruce took it in stride, having to keep his work secret from his family, who had moved to France to accommodate their son's ambitions. While working under Doctor Hoffman, Bruce spent his time at HEPA working on the computer software and the experimental physics aspects of the technologies.

The day Bruce learned of the existence of time travel was burned into his memory. Doctor Hoffman showed him in private in an experiment where a mouse was sent back in time thirty seconds with no visible side effects.

Thus, the Hoffman-Einstein Rosen Bridge (HERB) was introduced to Bruce:

On that fateful day, Doctor Hoffman approached a pimply-faced young Bruce Hayden while he was immersed in a complex algorithm of programming code; Doctor Hoffman indicated to Bruce to follow with a quick flick of his arm and a wide smirk on his face.

"You will play a vital role in the years to come, Champ," He put his arm around Bruce, his eyes glistening with pride. "And it will involve time travel." They both stared at the mouse walking around its cage, oblivious of its pioneering role in time travel.

Bruce and Leonardo did not comprehend the gravity of the prophetic statement until they had completed two more full years of work at the facility.

Suddenly, Bruce came back to the present, where Doctor Nault was staring at him and asking, "Well, what do you think, Mr. Hayden?" Bruce glanced at the board briefly and deciphered what Doctor Nault had been writing in chalk on the board, it was an introduction to the applications of integrals.

"Well," Bruce cleared his throat, a little embarrassed that he was daydreaming so thoroughly. "Well, I think that the integral of the equation that you've written and its corresponding graph are related in that the area under the curve you've drawn is equal to the integral of the curve's equation between those two points." Bruce stopped and briefly looked around at the other students who were looking at him.

"Very good, Mr. Hayden. I see you have been reading ahead in the textbook to the next chapter." doctor Nault smirked as he went back to the chalkboard and began explaining how Bruce had come to that conclusion and why the equation and its manipulation with Calculus had essential applications to engineering.

Bruce put his head down and flipped a few pages ahead in his text to get the class's eyes off him and back onto the lesson. He returned to the notes he had been taking and began to follow the study closely so he would not be caught off guard again.

I better focus, he thought. Allowing his mind to wander between future and past events was a sure way to lose concentration on his mission. He needed to be all business.

As the class ended, Bruce walked up to Doctor Nault and watched the day's lesson getting wiped off the chalkboard: the smell of the white chalk dust cloud sparked a litany of memories of working with Doctor Nault on orbital mechanics, nuclear technology, computer systems, and algorithms. All from another not-so-distant life.

Doctor Nault was staring at the board deep in thought as he wiped it clean with the large chalk brush as if he wanted to be sure to remember what he had written. This was not the case since the equations and corresponding graphs were elementary first-year calculus; the doctor's thoughts were elsewhere.

He and Bruce were alone, and Bruce watched the back of Doctor Nault's head as students' heads began peeking through the small rectangular window in the classroom door as they waited politely to come in. Doctor Nault was still staring at the chalkboard when he spoke.

"I was thinking, Bruce..." He fumbled with the chalk brush as he thought. "I ... I have been thinking a lot about what you said to me in my office the first day we met, and I was wondering if you would like to meet up regularly outside of classroom time, say on an evening or weekend, to come up with a working theory to integrate computing systems with quantum theory."

Bruce may have fallen over if he had not been close to the chalkboard. He regained his balance and smiled at Kevin Nault. AI did not predict the turn of events, even with his stellar start this time. He was dumbfounded at his luck.

Bruce stammered, "Uh..., of course, doctor Nault, I live within walking distance of the university." Bruce was contorting his face to hide his excitement, but his red-headed genetics gave him away again as his cheeks flushed severely. "So, whenever you want to meet and talk about those things and maybe do some computer programming trials, I would love that."

"That's great! Then it's settled." He put the chalk brush down on the chalkboard ledge and clasped his hands on both sides of Bruce's shoulders with a slap.

How about you come to my office this Saturday morning at eight for a couple of hours, and we will see how that goes."

More chalk dust billowed into the classroom as Doctor Nault put down the brush. He smiled at Bruce and opened the door, allowing Bruce to leave the room first. The freshly disturbed chalk billowed into the hallway as Kevin breezed out the door.

"Okay, see you then, doctor Nault," Bruce shouted with a wave as he walked past the students outside the room and down the hall; Doctor Nault had gone the other way.

Bruce quickened his pace, still in shock at what had just happened. He had some major restructuring to do and would have to re-examine multiple simulations when he got home; this was an absolute game-changer. He had to ensure that the good doctor's invitation would allow Bruce to steer him in the direction needed.

When he got home, he was sweating, partly from the pace of his walk and partly from the anticipation of how meeting with Doctor Nault weekly would change the simulation projections.

He gulped down a glass of tepid Willcox tap water as he sat at his desk and went to work.

Chapter 10: New Territory (11.09.1981)

The rest of the first week of school flew by for Bruce, who spent all his time outside school planning for his precious hours working with Doctor Nault. Early on Friday evening, Bruce was sitting in his basement study area examining a page of computer code when the holographic page disappeared. The lights dimmed, and the AI voice chimed. "Someone is approaching the door. It is Miss Rosie."

Bruce slid away his exposed tablet, walked up the stairs, grabbed a book and his ballcap at the top, and opened the front door.

"Oh, Miss Rosie!" he exclaimed as he walked out his door to meet her on the small porch. He turned and locked the front door behind him. "I was just going out for a walk before the Sun gets too low. I enjoy reading in the soft sunlight." He held up the book *20,000 Leagues Under the Sea* and smiled a big goofy smile. "What can I do for you?"

"Oh, well, uh…" Miss Rosie stammered briefly before composing herself. "How are you liking your accommodations so far?"

"I quite like the house," Bruce smiled. "And it's close enough to the university." Bruce kept walking, and after passing Miss Rosie on the porch, he stopped on the brown grass. "Is there something I can help you with?"

"Uh, no, I won't keep you from your walk. I just wanted to make sure your accommodations were satisfactory so far." Miss Rosie was stumbling on her words. Bruce had seen this behavior before; she needed money.

I'll talk to you later," Bruce yelled as he kept walking with his book open. He thoroughly enjoyed playing a clumsy, awkward, and socially inept teen. He kept walking to the sidewalk, pretending to be unaware of the not-so-subtle social cues Miss Rosie was giving him.

He glanced back several moments later and saw her going into her house.

Bruce closed his book once he was out of sight and then started to jog around to the back laneway of the house and through the edge of the wooded park, where he had made a discreet path in the long grass if he needed to avoid nosy neighbors. Bruce quickly squeezed through a space in the wooden fence that a conveniently missing board had created, and he could access the back of his house.

While standing above the basement window, he looked around him. Now completely hidden from Miss Rosie, he entered a sequence of numbers on his watch. There was a distinct 'click.' Bruce slid the basement window open, moved the blacked-out curtain, and shimmied his way into the window. Being skinny and tall certainly had its perks; he landed expertly on the basement floor, his long arm still grasping the window ledge.

After re-engaging the locking system on the window, Bruce slid the blackout curtain back in front of the window and leaned back in his chair. Tomorrow was his first out-of-school meeting with Doctor Nault, and after running dozens of simulations with all the relevant new data, he examined the results. Bruce opened his tablet, and the holographic video unit automatically began projecting the simulation results. He flipped his finger in the air, flipping through the pages to ensure he had a working plan entirely to memory.

This development had never happened before in his previous attempts, and he would not leave anything to chance with these extra meetings. He poured over quantum mechanics differential equations and quantum computing rudimentary theory. He also brisked through the Turing computer coding methods that he knew Doctor Nault was currently using and the computer simulation process to segway from Turing coding methods to the new computer coding they would use; this would allow them to start crunching volumes of asteroid data to calculate their orbits.

Bruce closed his tablet and put it away; the holographic image sat in the air momentarily before the light photons dispersed into the room's darkness. He sat a moment, with his eyes closed, focusing. He was ready.

Chapter 11: How to Blow a Prof's Mind 101 (12.09.1981)

Bruce's alarm went off as he was dressing; he had woken a few minutes earlier and had packed his bag, ready to go. Kevin Nault was an early riser, and he did his best work in the early mornings, often before the Sun had risen. Bruce grabbed his bag and held a warm piece of toast in his mouth as he opened the door to greet the fresh, cool breeze of the morning.

It was 7:30 as he walked past Miss Rosie's house, which was dark inside. She would not be vertical for at least a couple of hours. The singing birds and the soft morning sunlight made Bruce smile, and he inhaled deeply, enjoying the smells, both old and new memories, as they flitted through his mind. He was good in pressure situations, and this day could put him ahead in his mission by months, if not a year.

He thought of the possibilities as he started to run. Running always cleared Bruce's head, and he habitually ran to and from school daily. The two-kilometer run was a breeze for Bruce; he barely broke a sweat in the coolest part of the day, his long legs thrusting him quickly down the empty streets.

He got to the school earlier than the 8 o'clock time doctor Nault had given him, but he wanted to make a stellar impression today: not just a good impression, he planned to put on a clinic he had jokingly titled *how to blow your prof's mind 101*. He smiled as he opened the front doors to the Physics and Astronomy building where Doctor Nault's office was.

He ran up the three flights of stairs, taking the musty old air deep into his lungs, putting his hand on the old radiators sitting against the wall on the landing of each turn in the staircase. He reached the top of the stairs, stopped momentarily to focus, exhaled deeply, pushed open the third-floor door, and turned left towards Doctor Nault's office.

While walking down the long, polished green hallway floor towards Doctor Nault's office, he saw a familiar shape appear from behind a frosted glass office window. Kevin Nault opened the office door with a coffee mug in one hand, papers clutched under his arm, and, as usual, deep in thought as he closed the door behind him. He turned just as Bruce reached him.

"Ah, you're early, Bruce," Kevin exclaimed as he tried to hold the papers under his arm and moved his coffee mug to the other hand, thrusting out his right hand to shake Bruce's. Bruce shook his hand firmly.

"Yes, sir, I am ready to work. I'm excited for this opportunity to learn from you." Bruce was no stranger to playing to people's egos when necessary. "Well, I was just going down to the coffee maker for a top-up of my java. Why don't you settle yourself in my office? I'll be back in a couple of minutes."

Kevin started to walk down the hall, his whistling of "Dancing Queen" echoed through the deserted hallways; Bruce had many memories of long, pensive thought experiments with Kevin, many of them were accompanied by the radio music of the day in the background.

Bruce entered Kevin's office and saw a student desk pulled up to face the more prominent professor's desk. A pencil and paper lay already, and an empty chalkboard was ready for brainstorming. Bruce glanced over to Doctor

Nault's doctorate diploma, where he had left the nano camera his first day in the office; it would record and transmit audio and video to Bruce's tablet seamlessly, and Bruce would install two more, hopefully sooner than later. He would establish one of the cameras at the observatory where they would first discover NH-1983.

Bruce would install the other camera at Kevin Nault's house, where they would spend many hours working while the media squatted on the campus grounds to capture rare sightings of the famous Kevin Nault and Bruce Hayden.

He put his backpack on the desk and admired the camera's fortuitous placement to include the chalkboard and the two desks as Kevin came back in, no longer whistling.

"Good. You found your desk." Kevin smiled as he came in and shut the door.

"Yes, sir." Bruce sat down at the desk, grabbed the pencil, twirled it in his long fingers, and waited for further instruction from the professor. Kevin Nault went to the chalkboard, grabbed a nubbin of chalk from the ledge, and thought for a moment. Then, he wrote a title on the left-hand side of the three segments of the chalkboard: *Quantum Computing*. Half in thought, He dropped the chalk, realizing that the piece of chalk was too small to continue.

"So, Bruce," he paused, "I have barely stopped thinking about this quantum computing theory since the day we met, and I was hoping that we could brainstorm some ideas to develop a theory." He slowed his speech and trailed off at the end, staring quizzically at Bruce and fiddling with the new, longer piece of chalk he now held in his long, skinny fingers.

"So… how is it that an eighteen-year-old is sitting in my office, only having completed the first week of an undergraduate degree, *an undergraduate degree?*" he shook his head in disbelief as his hands raised up and then came back down, slapping the sides of his pants. He continued in his usual deliberate English cadence, "Is here to brainstorm with me about applications of quantum computing, the theory of Nobel Laureate." He threw up his hands again in disbelief at what he was saying.

Bruce thought for a moment. This had to appear to

be a collaboration; he needed to carefully let out the correct fishing lines to start reeling in this big fish. He cleared his throat. "Well, sir, I was only extending the quantum theory proposed in the Journal of Physics, and besides, it would be pretty neat to put that theory to the test, wouldn't it?"

Kevin Nault just stood there staring and smiling. "You bet your *ass* it would be neat" He pointed his chalk at Bruce with an emphasis on "*ass.*"

Aha! Bruce thought with a smile. Kevin hardly ever swore; if he did, it was only in the company of people he was comfortable with. Things were going just as Bruce wanted.

Bruce started to steer into a more focused path. "I was reading a book about Turing computer programming language, and I feel that even though it is a very good programming language, well, sir, it's just too brute force, and with the current speed of computer processors, it just won't do. I dream of being able to compute things that people think are impossible."

"Like what Bruce," Kevin hung on Bruce's every word. The long piece of chalk in his hands stopped moving.

The wheels were turning fast in the brilliant professor's mind.

Bruce continued with his well-scripted feeder lines. "Well, just for an example, a good application would be calculating the precise orbit of a near-Earth asteroid with the raw data from a telescope. And I don't mean by hand with a ledger; I mean a computer program that can accept raw data directly from the telescope, chunk through it at super high speed, and graph the points. Or accept raw data from any physics experiment, for that matter, and process it quicker than ever before."

Better leave out the part that humanity would soon be able to accept raw data from a remote location, wirelessly, anywhere in the world or space, Bruce thought.

"Do you mean to enhance the Turing programming language to make it faster computationally?" There was a puzzled look on his face; Kevin was no slouch. The father of modern computers, Alan Turing, had his name attached to the programming language that would rapidly become obsolete in the breaking dawn of a supercomputer era that would see exponentially increasing computing speeds for an entire century.

"No, sir," Bruce was confident. *Prepare to have your mind blown, Kevin.* Bruce thought for a moment. "Sir, I propose that there is a faster way to analyze complex mathematical problems."

Bruce paused to get this exactly right. Kevin was already sitting motionless, mouth gaping open; the chalk was hanging precariously between his index and middle fingers, like an old drunk about to drop his cigarette. "Sir, what do you think about brainstorming to create a more efficient computational language that could accept a huge amount of data?"

Bruce leaned back in his chair as he watched Kevin Nault absorb what had just been said by the eighteen-year-old sitting in front of him. The two just stared at each other. Kevin rubbed his hand over his face and head, indicating that he was overwhelmed by the possibilities and potential of this young man from Canada.

"Okay, Bruce," Doctor Nault brushed away the quantum computing title and stared at the board. He paused and shook his head as if that would slow his racing mind. Bruce knew he was trying to think of the title for this brainstorming session for a more efficient computing language.

Bruce interrupted the professor's pause. "How about we start with brainstorm A."

Kevin Nault robotically wrote the words on the chalkboard, nodding as he thought about what was happening. Physicists were used to using variables in their work, and A was as good a place to start as any. More comfortable than the modern English alphabet, physicists traditionally used Greek symbols like alpha and beta, but Bruce needed them to settle on C eventually since the C programming language was due to be created in a couple of years and not changing too much about the past was an essential factor in Bruce's mission.

The programming language Bruce and Doctor Nault would develop just half of the computational problem.

The other half of the problem was developing the working theory for quantum computing. Bruce and Kevin's creation of the idea was inconsequential to the mission since Bruce would embed the quantum computing hardware into the prototype without anyone's knowledge.

He needed Kevin to be a part of this process, which had to be believable.

Bruce already possessed the computing hardware they needed, which he had in the tablet in his workspace in the basement of the house. They were going to spend a lot of time developing the theory just for Doctor Nault to feel that the things that were happening were plausible; he needed Doctor Nault's undivided attention for several years for this work to take flight, and that meant involving him in all necessary facets of the mission. That is all the parts Kevin *could* know about.

In a seemingly long-ago life, in desperation, Bruce made the catastrophic mistake of telling Kevin the *whole* truth about the mission, contributing to its failure.

Bruce watched as Kevin wrote some initial brainstorming thoughts on the board. He chimed in a few times to keep the doctor moving along, but he needed to play the student and delicately guide the professor.

The two-hour meeting between Bruce and Kevin ended up being a polite chess match. Kevin tried to steer towards quantum computing theory while Bruce drove towards the new programming language. The result was a dead-end discussion in which Kevin was determined to marry the square peg of quantum mechanics into the round hole of the Turing programming language.

"Well, that was a great discussion." Kevin was exasperated but excited about their discussion.

"Yes, it was, professor," Bruce replied, "And thank you for spending this time with me. Can I assume we will meet again next Saturday at the same time?" Bruce knew the professor was a creature of habit and was now completely invested—*hook, line, and sinker*.

"Absolutely, Bruce! Next Saturday morning at eight, it is." Kevin tossed the chalk onto the lip of the chalkboard and wiped his yellow and white chalk-dusted hand on his sweater vest. "See you in class, Bruce."

"Yes, sir." Bruce walked out of Doctor Nault's office, and his head was down while he walked to the staircase, down the stairs, and into the distinctly hotter outside air. Bruce would have to do his homework and pour through multiple simulations before their next Saturday meeting.

Kevin would come around to where Bruce needed him to be, but he was extremely stubborn and would need some strategically placed lures to get him there. Bruce was interested in what the AI program came up with:

It had been silently watching the Saturday morning meeting from Bruce's basement, its eyes and ears perched to the side of Doctor Nault's doctorate diploma: observing, predicting, extrapolating.

No sound came from the basement equipment that sat in the dark on Fremont Street; it was the silent watcher.

The 22nd-century AI technology watched, recorded, transcribed, and processed all the data from the two-hour meeting while Bruce was walking home: speech pitch, cadence, body language, thermal temperature graphs, and heart and breathing rates. The AI overlooked nothing.

By the time Bruce got home, a report was waiting for him. He sat in his chair, and with the flick of his finger, the holographic projector silently started, adding an audio and visual component to the previously dark space. AI presented the highly analyzed data efficiently so that Bruce could absorb the information.

As Bruce watched the AI presentation, he watched Doctor Nault's live image at the bottom left of the holographic data as he made his handwritten notes from the blackboard brainstorm.

Bruce occasionally glanced down at Kevin's video feed as he sifted through all the day's data. He was excited at the thought of the flawless start to the mission.

Chapter 12: The Future is Calling: (08.03.2135)

"I will not yield!" His tenor voice and ever so slight French accent reverberated through the HEPA laboratory. Everyone stopped in their tracks. The conversation had just ended as Doctor Gus Maxwell stormed out of the lab, and the sound of the door pistons hissed as they closed, invading the sudden silence.

Doctor Leonardo Hoffman stared at the other two scientists who remained in the lab; his face looked like a purple tomato, his heart was racing, and his eyes looked wild. *Who were they to make such a drastic change to this mission? My life's work?*

Doctor Fran O'Shea and Doctor Mitch Smith were now looking at the ground. Mitch spoke first while Fran leaned on a tower of computer processors and servers as she listened.

"Leo, look," he sighed.

"Gus was too forceful, but we are scientists; we must consider his input, otherwise…" He paused and looked at Fran, searching for the right words. Confrontation was not his forte. He looked back at the ground.

"We need to do this together," Fran added in her calming but assertive tone.

Doctor Hoffman was still angry.

"He has no right to demand that we change this mission after years of planning and *decades* of my work…. our work." He took a deep breath and exhaled, his cheeks puffing out with his breath as a bead of sweat ran down his face. He deflated somewhat as he thought about their work as a team over the years. I'm sorry for my outburst."

Fran, the peacemaker, spoke again. "Leo, we are all stressed out. We have all been working way too many hours lately. Gus didn't mean anything other than to speak openly about other possibilities for this mission."

"I know, I know." Leonardo was embarrassed and became subdued from his previous stance. He took a moment to compose himself before he spoke. "We agreed, and the board of directors has agreed, that I am heading this team, and just because I am the only one in this lab who has the mission security codes does not mean I won't hear what everyone has to say. We are a team, but I cannot compromise the mission in this manner. I will not allow it!"

Fran looked at Mitch, raising her eyebrows at him. Mitch looked at her, then back down at the ground, and nodded, conceding his confrontational weakness. *Great, he's got his head in the sand again.* Fran glared at Mitch. His gaze never left the floor as she stared at the bald spot on his head. His round frier Tuck-style bald spot on the top of his head might have burst into flames if Fran's gaze were more intense.

Fran continued. "Leo, Gus is not asking you to compromise this mission – we are ..." she glared over at Mitch's bald spot again that was still facing her, his eyes still transfixed on the floor tile in front of him. "Asking you to consider taking this precaution for the sole reason to ensure success this time."

"So, you think we should talk about this?" Leonardo felt the change in the room as he looked back and forth between the two scientists. *They were all in on this*!

Mitch sat up in his chair and then, with his best confident look, stood up next to Fran. "Yes, Leo, please consider how we can ensure success. What if Bruce fails again?" His voice crackled from his unexpectedly dry mouth. He swallowed nervously.

"He will not fail!" Leonardo Hoffman stood up firmly, pounding the desk with his fist as he stood; he had absolute faith in Bruce Hayden. "Adding any parameters to this mission will only increase the volatility and the probability of failure."

He left the room after he spoke, and Leonardo did not look back as he walked out of the HEPA lab and left the area designated "Level 3 Access – HEPA". He nodded at the guard standing between the two sets of doors beside the sign as he brisked past her, and Leo headed down the hallway towards Gus' office, where he figured Gus had gone after his outburst.

He needed to do damage control.

His team worked well and at a consistently high level; what he had just seen deeply concerned him. It was a scientist's version of a mutiny. Of course, he never told his team that there was one other person with the mission codes, nor would he. It would be a foolish proposition to any scientific panel to have just one person controlling access to mission parameters while the mission was ongoing.

Being the team leader, he had extra responsibilities, and since he created the HEPA technology, he had extraordinary clout with the board of directors. He had insisted to the board that Bruce be chosen for the mission one last time, even though the board members had placed Bruce in a tie with another candidate. Doctor Leonardo Hoffman's vote was the deciding vote, and even though the board had questioned him about his sentimental attachment to the young man, he insisted that Bruce would be successful this time. *Bruce could not fail.*

Chapter 13: Time Flies, when you're Racing it (18.11.1981)

The following two months saw Bruce attending classes during the day and spending most of the nights under the blue holographic lights of his basement. His hand-drawn blueprints for the asteroid lander craft and landing pods were about 5% complete, which would have been a much more remarkable feat if he had not re-drawn them from his previous blueprints stored in his holographic unit.

As he re-drew the schematics, he refreshed himself on all aspects of the asteroid lander unit and pods that he and Doctor Nault would build with a small crew of mechanical and aeronautical engineers. He drew the schematics on large paper he had gotten from the university, and as he drew, the holographic unit was saving the drawing. After completion, he would have to tear up and burn the papers; keeping the significant rolled-up documents in his basement this early in the mission was a risk he would not take. Nothing would be left up to chance this time.

His weekly Saturday morning meetings with Doctor Nault were progressing nicely, and Bruce was working on broaching the topic of joining Kevin at the observatory for his Tuesday or Saturday night telescope time at the Steward Observatory outside of Tucson. Doctor Nault was nothing if not dedicated; he spent his daytime hours during the week teaching at the Physics and Astronomy department, answering student questions, and supervising research projects and experiments in the physics lab. Kevin's singular focus on the computing theories he and Bruce were developing would drive their progress.

After their meetings, Kevin spent many hours writing and reviewing his thoughts in his scientific journal.

Bruce was aware of the doctor's habit and had watched Kevin spend many hours drawing diagrams and writing thoughts and formulae in his journal. Bruce did not attempt to look in the journal because he felt he knew what was in it, as he gleaned glimpses of it periodically from the camera installed on the doctor's diploma.

The photos of Kevin's notebook captured by AI were stored and incorporated into AI's calculations.

Kevin often wrote on the chalkboard from his small black hardcover notebook when Bruce arrived for their Saturday morning research collaboration. The blue ink stain on his left hand from the fountain pen he loved to use was a tell-tale sign that Kevin had been delving into his mind before their meetings.

The glass ink container and three different writing tips always sat perfectly on Kevin's desk to the left of his workspace. However, the ink was never left even close to empty. The university had gotten into the habit of keeping the professor's favorite blue ink in the stationery room as an ongoing thank you for the brilliant professor's teaching and scientific research that he was producing in the name of the university.

Kevin never showed Bruce the notebook; he always put it back in his desk or shirt pocket after copying formulae from it onto the chalkboard. They never spoke of it, and, for the time being, Bruce was not acknowledging he knew about the notebook.

When he was not working at the university, doctor Nault spent every Tuesday and Saturday night at the Steward Observatory researching Near Earth Objects. He always looked tired when he came to teach at the experimental physics lab on Wednesday mornings, and for good reason: he religiously spent his time from 9 p.m. to 3 a.m. at the observatory on Tuesday nights. He did not waste a moment of his precious time procured by the university to do his research.

His well-established routine, however, would be turned onto its head when he and Bruce made their historical discovery of the asteroid NH-1983. After the discovery of NH-1983, the Steward Observatory telescope would essentially be theirs to use whenever they wanted. However, after their discovery, they would spend considerable time at the Yerkes Observatory in Williams Bay, Wisconsin. Williams Bay was a nice, quiet place to conduct their observations and research without being suffocated by the constant pressure from the media.

The lander prototypes and pods would be built secretly in the ample open indoor space at the Yerkes Observatory. For good reason, their location in Williams

Bay would be hidden from the public.

Bruce completed the schematic drawing of the propulsion system of the landing unit he had been working on, and the holographic projector dimmed as the room background lights increased in intensity, revealing the simple basement space. Bruce stood and stretched through his quick body-energizing and brain-stimulating exercises: breathing, jumping jacks, burpees, push-ups, and sit-ups.

When he finished, he sipped his tea, which he realized was cold. He put the cup down on the third bottom step behind him. The clink of the coffee mug hitting the old wooden step echoed briefly in the basement. Time flew by when he worked, mainly because Bruce had a remarkable focal attention span. His security system at the house pulled him out of his concentrated state so that he would not be surprised by Miss Rosie or any other person who came to the house, even though that was rare.

Thinking about Miss Rosie, Bruce looked at his watch and saw it was close to the end of the month. He had blown her off the last time she had come over needing money, and that was several weeks ago. He ran up to the bedroom and put together a modest amount of money, put an elastic around it, and put it in the top drawer of his dresser.

Miss Rosie played a small part in Bruce's plan, but she was part of his plan nonetheless. After his first mission, he brought Miss Rosie bucks, which he had labeled the acronym MRB, onto the money destined for her.

Soon, he would be on a salary paid by the university, and money would not be a problem for him. He felt that Miss Rosie should benefit from her unknown role in the mission. Shortly, the university was going to be paying Bruce a student salary and offer him a full scholarship after the discovery of the asteroid NH-1983 so that they could ensure his student enrolment status there. He and Doctor Nault would bring plenty of media coverage to the university, creating a boon for research, physics, and engineering programs that would last for decades.

Bruce brought the modest bundle of Miss Rosie Bucks down the stairs and into the basement. He flopped the money onto his desk and took a deep breath as he started pacing back and forth in the basement space. Looking ahead into the mission was a guilty habit that Bruce had become accustomed to; he never saw the harm in reviewing a few steps forward so he would be able to prepare appropriately.

It was perplexing to Bruce how events in time happened as if in a pre-conceived notion, similar to water flowing along the same contour. If the water or contour were not disturbed, the water would travel the same path, just like history followed the same direction as Bruce re-lived moments from two previous missions.

He ran his hand through his hair and came back to the present. He had more work to do that night.

Chapter 14: Something's Horribly Wrong (08.03.2135 08:23 hrs.)

"Give me the codes." The masked gunman's eyes revealed his desperation to Doctor Hoffman. The gunman's voice was electronically altered.

Leonardo's mind was racing.

"I...I don't know the codes," Leo stammered. He was on the ground; his head throbbed from being pistol-whipped on the side of his face. He was trying to buy time; he needed time to think.

How did he get into this secure area?

As Doctor Hoffman's eyes darted around the entrance to Doctor Gus Maxwell's office, his heart stopped. He saw Gus's unmistakable long boat-like brown loafers sticking out past the back of his desk. A pool of blood came out from behind the desk, almost to the doctor's large shoes.

"You bastard!" he yelled at the gunman as he winced in pain.

The gunman slowly pulled back the cocking lever on the handgun. Looking down the barrel of the pistol, Leonardo Hoffman, always a keen observer, noticed something unique about the gun's appearance.

Although made to look like metal, the porous surface appeared cracked on the handle, likely from striking Leonardo's head. The gunman's pinky finger was not entirely around the hand grip, and Leo watched the end of the pinky finger shake slightly back and forth as time slowed down to a crawl. Leo's attention went back to the end of the barrel still pointed at him, and he examined the porous surface of the handgun's frame.

The gun and bullets must be made of some synthetic, was the last thought that flitted through his mind before he felt as though he was hit with a sledgehammer in the left shoulder and knocked onto the floor in searing pain.

"Okay, okay!" he yelled, his right hand up towards his attacker was open. "Okay, I'll tell you the codes!"

Damn those huge doors to the lab, he thought. *They*

won't hear me yelling, and that gun has some kind of silencer.

"Help me up!" He glared at the gunman, "How are we going to get to my office with my shoulder and head bleeding?"

The gunman said nothing but opened a small packet of white powder and messily emptied it on Leonardo's shoulder and the side of his head.

"Rub it in," the highly digitized voice commanded.

The doctor used his right hand and gingerly rubbed the gritty white powder on the left side of his head. Then, he did the same on his left shoulder. The pain in his head and shoulder decreased dramatically, and he sighed with relief.

"Don't confuse your lack of pain for my weakness, doctor," he croaked. "I will put the next bullet between your eyes, and no amount of powder will help you with that. The digitized voice had a vaguely familiar cadence, although Leonardo's spinning thoughts could not place it.

Doctor Hoffman slowly rose and left his friend's lifeless body behind the desk. The gunman closed the door to the office behind them, and he followed Leonardo farther down the hall toward his office.

Nobody will be down here, and no cameras broadcast in this secure area, he thought as he walked farther down the hallway toward his office.

Think Leonardo, think!

He got to his office door and opened it. He glanced around frantically, thinking of what to do. *Those codes could not be compromised at any cost.* He sat at his desk chair and hovered his hands over the virtual keyboard. The gunman came in and closed and locked the door with the deadbolt.

There was no window except the frosted glass window on the top half of the door, reinforced with a polycarbonate mesh. *There was no way out.* The soundproofing of offices, the no cameras policy, and the high-security grade doors had been Doctor Hoffman's idea in the construction phase of project HEPA offices. If he did not fashion a plan expeditiously, it would cost him his life.

"The codes," the digital voice ordered.

"Okay, just let me get onto the computer." Doctor

Hoffman was sweating profusely, and his shoulder started to throb again.

"They are complex codes that change every twelve hours. You have to be patient."

He began typing on the virtual keyboard with only his right hand. His left arm dangled down beside him in the chair. Leo typed the long sequence of characters with awkward one-finger air keystrokes.

"Any tricks, and you will die." The assailant pointed the gun at Leonardo's head; another non-metallic round had been auto-loaded into its chamber.

"If you sabotage this mission, we will all be destroyed." Leonardo was in pain as he looked his assailant in the eyes.

Doctor Hoffman was stalling.

"That's the idea." the gunman grumbled. His digital voice changer was faltering, and the electronic voice glitched. He began to move behind the desk with Doctor Hoffman.

"No!" Leonardo yelled with his uninjured arm

outstretched to his assailant. "This password system uses facial recognition along with fingerprint and retinal recognition. If more than one person looks at the screen, it will not work!"

The gunman took an impatient step back and stood off the side of the desk, pointing the handgun at Doctor Hoffman's head. The end of the gun barrel touched Leo's right temple. It was still warm from its last shot. "I'm waiting," croaked the gunman.

Just a few more seconds, Leonardo thought as he tried to manipulate the object in his left hand. His whole arm was almost numb and throbbing with pain. He took a deep, shaky breath. *This is how it had to happen; this was his only course of action.*

"Please don't shoot her!" doctor Hoffman yelled as he looked at the frosted glass door to the office.

The gunman's attention was moved from the doctor for a matter of a second as he pointed the handgun at the doorway, and, in the time it took him to look over at the door and realize that he had been fooled, there was a brilliant flash of blue light as the gunman looked over where the doctor was sitting. He rushed his arm back to point the gun at Doctor Hoffman and fired into the brilliant blue light that contained the doctor's silhouette.

Doctor Leonardo Hoffman had vanished.

Chapter 15: Blast from the Future (21.11.1981)

It took Bruce a matter of minutes to prepare for his unexpected early morning trip into the desert and three hours on foot to reach the location given to him as the scene of emergency transport. Upon arriving in 1981, the second time travel orb emitted a high-frequency emergency signal that only another time travel orb could detect. The only person who knew the device's minutia better than Bruce was Doctor Hoffman.

Bruce was working in his basement when he was alerted to the problem; he knew it was not good news as he packed his bag to venture out into the desert sands. His training kicked in as he filled the several small bags containing a first aid kit, food rations, emergency blanket, carbon fiber poles, fabric, enriched water, and intravenous saline bags. Bruce packed all these things calmly and orderly, as he had been trained to do many times. Anxiety was useless to him as he attempted to conserve as much energy as possible during the trip into the desert.

Bruce had been made aware of the second HERB time travel orb before his missions, but its existence was known only to a small core group of people. As the morning sun glanced over the horizon, Bruce alternated between running and walking to conserve energy. He had brought as much water and first aid supplies as he thought he could carry efficiently, but he packed light, not knowing the physical condition of the time traveler who was now alone, at sunrise, in the desert.

He had already tried to communicate with the other orb's user, but there was no response. The initial visual image transmitted was only of a dimly lit sky with a cactus arm off to the side in the distance. The orb was sitting in the sand, or an open hand, but no part of the image showed a person. Then it went black. The sphere went offline.

Bruce stared at the black image and thought of all the possibilities as to why he could not communicate or see the impression given by the second HERB device: a power source discharge, an unknown malfunction, or a manual shut-off.

Going over the facts he knew about the twin time travel devices, Bruce hoped for the best but prepared for the worst. The quantum entanglement of the two orbs meant that they could communicate through space and time; whatever action one of the devices had completed instantly, the other orb knew of it, regardless of distance, space, or time. Albert Einstein called quantum entanglement "spooky action at a distance" because it seemed to defy the laws of the physical world.

The evolution of quantum mechanics and the invention of time travel in the 22nd century satisfied the laws of the physical and quantum worlds.

Bruce's mind wandered through all the possibilities during his long trek; *if the orb was activated and the settings unaltered, the person would be brought to the exact time and place as the twin orb's traveler.* That previous time traveler was Bruce. Since Bruce had been alone in 1981 for almost four months, he knew that the time settings had been altered.

The unknown traveler must have altered the destination time settings and, therefore, had some knowledge of the device.

Then, the device just stopped communicating.

Bruce tried again in desperation to communicate with whoever had just arrived in 1981.

"This is Alpha-Gamma-1 transmitting to Beta-Gamma-2. Come in Beta-Gamma-2."

Nothing.

Bruce tried to do a second remote scan of the second orb's surroundings. There was no response; it was still offline. He was still several kilometers from the device's last location. He increased his pace.

Chapter 16: Doctor Hoffman's Awakening

He could see just a faint light as he fought to open his eyes. One eye opened and saw the strange, blurry surroundings; the other eye was swollen shut. His left eye was warm, and he could feel the blood rushing to his throbbing eye.

Hesitantly, he moved his right hand and touched the left side of his face; he flinched. The deep ache in his head seemed to accompany his badly swollen eye. Memories of a distant time started to come to him, but they were blurred and distorted, and then they stopped.

Just pain and confusion.

He tentatively moved his right hand from his throbbing eye and let it rest on his left shoulder. Although he could feel it was wet on his hand, he could not feel his hand touching his shoulder. His whole left arm had no feeling. He looked momentarily at the sun on the horizon: *was it rising or setting*? The soft sunlight rays were refracting through the tears in his eye, disorienting him. He tried to turn his head to the left so that his open eye could assess what was wrong with his left arm. The searing pain prevented him from moving his head more than ten degrees to the left. He winced and straightened his head.

Where the hell am I?

He took some long, labored breaths and tried to compose himself. He was obviously in shock. Lying on the side of a small but steep sandy hill, his feet facing downhill, he could see a dark black sphere whose smooth trail in the sand revealed how it had gotten there. Leo stared at the globe; he knew it well, yet he could not think of what it was or why he would have it there. All he knew was that the sphere was critically important, and he tried to sit up to get down to it. Leonardo used gravity to his advantage and leaned forward as he struggled to get on his knees. The pain was unbearable, and he whimpered before he saw the darkness closing in from the sides of his vision in his good eye. He fell forward down the hill and lost consciousness moments before he landed on the orb, covering it from sight.

Chapter 17: Doctor Gus and the Bullet Hole (08.03.2135 10:07 hrs.).

"I don't know how that blood got there!" Fran was indignant as she pointed to a small pool of blood on the floor at the entrance to Gus' office.

"How could someone get in here to do this to Gus? And Doctor Hoffman just vanished from the lab?" She was sweating copiously. The police officer was staring at her. She never had a panic attack, but this seemed a good reason to start.

"Doctor O'Shea, I need to get a statement from you about the last time you saw Doctor Hoffman or the last time you saw Doctor Maxwell alive. Were there any conflicts? Is there anyone you can think of who would do this? I'll be asking you that kind of thing." The police officer put his hand on her shoulder. "Why don't you come sit down in the hallway?"

"How do I even know I can trust you?" She

mumbled, deep in anxious thought.

How the hell could this happen? she thought. She felt the panic move up into her throat again.

"There has been a major security breach in this lab, sir." She looked at the officer intently; he seemed young. "Our lead scientist is missing, and another is dead. You don't understand how important it is for this entire lab to be processed forensically to discover what happened. You say that Doctor Hoffman didn't leave through security?"

"No, doctor O'Shea, the last entries in the security footage confirm independently and show the four of you coming into the lab area this morning, and that's it before you called us here. Nothing else."

She sat on the floor, her head spinning. Gus was lying dead in his office in a pool of blood, and Leo was missing *vanished.*

Oh, Leo, I hope we didn't send you over the deep end. I can't imagine you doing something like this.

"And you said the entire HEPA lab area, including CERN, is on lockdown?" she asked the officer; her thoughts were running wild.

"Yes, ma'am, the whole place is locked down; no one exits or enters until we sweep this entire facility. Given the ultra-high security of this particular area, we would like you to stay to observe the forensic officer while he documents all of the evidence." She nodded, her head resting on her knees as she sat on the floor.

"Is Doctor Smith okay?" She asked, her voice muffled by her knees. She tried to hide her sarcasm; *he's such a boob,* she thought.

"Yes, he is still in the medical wing of the facility being treated for shock." The officer squatted down in front of her. His tone became more compassionate. "Are you okay to stay to allow access to the forensic officer?"

That wimp left me here to deal with all of this, she thought. She nodded at the officer, took a shaky, deep breath, and raised her head to look at the officer. "I need to go and see Doctor Hoffman's office again."

The officer briefly thought before he spoke, "Okay, but you can't touch anything. We have multiple nanocams installed throughout this area now so that we have continuity of this scene and don't miss any evidence."

She nodded as she got up, holding onto the wall in case she was still dizzy. She stood upright and started walking towards Doctor Hoffman's office. Suddenly, she stopped and stared. A dark red blood droplet was on the floor, then another a short distance away. It had been just over an hour since Fran discovered Doctor Maxwell's body in his office, but the blood droplets in the hallway had already started to coagulate and dry. She pointed them out to the detective with her extended arm as she took a wide berth, covered her mouth, and continued towards Doctor Hoffman's office.

This has to be Leo's blood, Fran thought. *There is so much blood around Gus' body he couldn't have run up the hallway.*

She looked at the floor before she walked into Leonardo's office. *No blood pool, thankfully.* She scanned the area where his empty chair and desk drawers were. His computer screen was dormant, as was the processor. Her eyes scanned the desk and chair for blood drops and saw something on the wall. She froze.

There on the wall beside Leo's chair appeared to be a bullet hole in the wall. *Oh my God, Leonardo, what happened?* She backed away from the desk and back into the hallway. She was sweating, the panic felt well above her throat, and she started to get dizzy again. She stood there, her thoughts spinning around; this was a catastrophe, no matter how she looked at it. *Was the mission compromised*? She continued to stare into the open door to Doctor Hoffman's office.

Leo, what happened to you? What happened?

Her thought was interrupted by the officer, who cleared his throat and spoke a second time.

"Ma'am, your phone." The officer was pointing to the buzzing phone in the pocket of her pants.

She fumbled with the phone as she took it out of her pocket and slowly raised it to her ear; it was the director. She stared at the phone screen as she put it up to her ear. *Gus is dead, and there's a bullet hole in the wall in Leo's office. And Leo's missing.*

She took a deep breath. "This is Doctor Fran O'Shea."

She had a lot of explaining to do with little information.

Chapter 18: The Rescue of Doctor Hoffman

Bruce sprinted the last few hundred meters towards the motionless person who lay face down on the hillside. As he got closer, he saw the salt and pepper wispy hair at the back of the person's head.

"No!" he yelled. *It can't be Leo. Please let it not be Leo.* Bruce's head was spinning. *What could have gone so wrong?*

Bruce skidded to the doctor's side with a wave of red sand coming to rest against his lifeless body. Bruce rolled him over; what he saw frightened him badly.

"Doctor Hoffman! Leo!" Bruce shouted.

He was breathing. He took a bottle of water from his bag and poured some over the sand caked with blood on the left side of Doctor Hoffman's face. *His left eye was swollen shut, and a large contusion had started to form on the left side of his head near his temple.*

Bruce took out his first-aid scanning device and watched it scan the doctor's body. He read the result of the scan on the screen:

Heart rate: 168 bmp

Blood pressure: High: 170/120

Left eye – large contusion / laceration

Left temple – large contusion

Left shoulder – damaged, foreign body lodged in the shoulder; possible gunshot wound.

Bruce read the screen in disbelief. He pressed a few buttons on the first aid unit's screen:

Action: Minimize bleeding, stabilize heart rate and blood pressure, and treat for shock.

Bruce began to work on Doctor Hoffman, and as the doctor moved in and out of consciousness, he moaned in pain, and Bruce could see the disorientation in his one functioning eye when it opened.

He needed to work fast.

Doctor Hoffman opened his right eye and looked around. Confused for a moment, the last events he could remember came flooding back to him, and he sat up and felt his head hit a covering of some kind. Then, his shoulder and head throbbing forced him to lie down again. He moved slightly from his flat position and tried to take in his surroundings.

Am I dead? What is this place? I must be hallucinating. His thoughts raced as his eyes teared up; the emotional toll of what had happened seemingly seconds ago started to overflow.

He could see a sand-colored covering over the top of him, and although he couldn't see his left shoulder, he realized there was some dressing or bandage on it. He couldn't hear anything but a high-pitched ring in his left ear. He struggled to listen to what was around him with his right ear; the wind, the covering over the top of him was fluttering, and the unfamiliar cry from a bird flying overhead.

Leonardo attempted to roll onto his right side so he didn't have to put weight on his left shoulder. He grimaced as he rolled but made it onto his right side. He took in the different angle of sight: an intravenous tubing that led from a first aid kit to his right hand, a thin black metallic structure that was holding up the covering over the top of him, and lots of bloody gauze. From this angle, the first aid kit was mostly empty, save for the tubing coming out of the equipment that led to his arm. He was receiving fluids via this intravenous line; he could hear the faint whirr of the machine pump every few seconds now that his good ear was close to the ground. He heard footsteps approaching and a dragging sound.

"Help!" he yelled in a dry, hoarse voice. The fabric covering Doctor Hoffman lifted off the ground, allowing a familiar face to enter.

"Hey, Leo. I'm glad to see you awake!" The face came into focus. "Do you know who I am or where you are?"

"Bruce?" Leo thought for several seconds as his good eye teared up again. *The masked gunman, getting shot, using the time travel device from his office chair.* "Oh shit... I'm so sorry, Bruce, I had to do it." He paused as the reality became more apparent, "I'm in the 1980s, aren't I?"

The face moved in close; the boy had a distinct look of worry. "1981, Leo. What the hell happened? Who did this to you?"

"I…. I don't know. I was attacked by a man holding a gun. His voice was electronically-" Leonardo coughed. He could feel his scratchy, dry throat, mouth, and lips stinging.

Bruce opened a water canister and put it in Doctor Hoffman's mouth.

"Take a sip. It's hot out here." He sipped the water as best he could, given his sideways position. He chose his words carefully so he didn't have to talk too much; he was exhausted. Doctor Hoffman spoke in a deliberate tone in broken English.

"Voice altered, wanted mission codes, shot me, I used the twin device." He slowed, his voice trembling as he spoke, "Where is the HERB device?" doctor Hoffman began to feel panic welling up inside him when he didn't get a direct answer.

"Here, let me help you lay down; you're tired and need to rest. You've lost a lot of blood. We have a long way to go tonight after the sun goes down." Bruce's voice was shaking badly. He aided Doctor Hoffman to lay flat on his back and watched his eye slowly close. He was asleep again.

Bruce got out of the makeshift tent and got to work. He had a couple of hours before the sun would be low enough to start back to town with the doctor, and he had to construct a stretcher to drag his patient across the desert. He was going to use the carbon fiber poles he had used to hold up the sun shield fabric covering Doctor Hoffman and weave the poles through the eye holes in the material.

There was no risk of the fabric ripping or getting damaged even if Bruce dragged it all night; it was a rugged 22nd-century material with an enormous tensile strength; different materials were woven into the fabric, depending on the intended use. The fabric Bruce was using was super strong but also had sun-reflecting properties and heat-sensing nano-nodules; these nodules would let the proper amount of heat through to make it comfortable for a person underneath and reflect the rest of the heat into its surroundings.

Bruce worked briskly to construct the mobile stretcher to start moving Leo when the sun was low enough to keep the doctor warm but not too high to raise his body temperature. He had often constructed the emergency stretcher in training but never during a mission.

For Bruce's mission, nothing was overlooked; four 25 cm diameter wheels could be constructed and attached to the stretcher so Bruce could pull the stretcher on smoother surfaces and lift and pull the stretcher on its back wheels if he desired. Bruce would have to use both methods to get the doctor back to his house under cover of darkness.

As he left, pulling the sleeping doctor Leonardo Hoffman on the stretcher, he was still reeling from his earlier discovery. *The second time travel device was missing.*

The sun was taking its last peek above the desert horizon when the man with the backpack started his journey, pulling the loaded stretcher behind him. His silhouette silently danced along the darkened desert sands, expertly weaving through the rocky outcrops and amongst the cacti. The body in the stretcher lay motionless. The animals and insects of the night who were waking watched the unusual sight; their silenced calls were the only indication that two time-travelers who had reunited under the desert sun that day were on the move.

Chapter 19: Damage Control (08.03.2135 12:15 hrs.)

Doctor Fran O'Shea sat precariously on one of the HEPA lab chairs in the spacious industrial room where they conducted experiments and held meetings. The forensic officer was still methodically moving from room to room, scanning for evidence; a small drone was hovering behind him, recording the events and all the observations of importance.

Equipped with a video camera that could see regular wavelengths, infrared, and other precise wavelengths of light, the drone could see bodily fluids and gunpowder residue. Fran's right knee bounced up and down at a furious pace as she waited. Her conversation with the director from earlier was still ringing in her ears:

"So, walk me through this again, Fran; there is blood around Gus' body in his office and another pool near the door to his office. There are blood drops in the hallway and a bullet hole in the wall in Leo's office."

The director's voice was calm and methodical.

"Yes, director." Fran felt like she was teetering on the verge of a nervous breakdown.

"Fran, go over to Leo's desk." Fran walked up the hallway past the forensic officer; he had already processed and documented Leo's office.

"Ok, I'm in Leo's office." Fran expected the director to ask for another visual description, but his following words were unexpected.

"Close the door and lock it." Fran slowly went to the doorway, peeking out into the hallway. Not seeing the forensic officer, she quietly closed the door and slowly engaged the deadbolt.

"Now, go to Leo's desk, the middle drawer to the left of his chair." Fran followed the directions and stopped in front of the desk; the bullet hole in the wall was to her left.

"Ok, I can see the drawer," she said.

"Put the phone up directly in front of the second drawer." Fran was confused but followed the directions.

"Ok, it's in front of the drawer."

"Stand up and hold the phone with your arm extended so it is in front of the drawer."

As Fran stood up, holding the phone, she could hear the director's voice giving some voice command code, and the handle on the drawer lit up a faint orange color. *What the?* Fran's heart raced. The drawer gave a single barely audible beep and then opened.

"What do you see in the drawer?" The director's voice was still cold and monotone. Fran stepped back closer to the desk to the left of the chair and looked in the drawer that had opened with the voice command from the director.

"Nothing. There's nothing." Fran's confusion was mixing with anxiety.

The director yelled, "Check the damn drawer! What's in the drawer!?"

Fran frantically put her hand inside and swept it around inside the drawer—nothing but a smallholder of some sort.

"Nothing, Nothing! Nothing is in the drawer except some little holder with nothing in it."

The director's cold voice continued back in its regular slow monotone cadence, giving Fran goosebumps.

"Do nothing, do not leave the lab, and tell no one about this conversation until you hear otherwise from me. Do you understand?"

"Yes...." The director hung up before Fran could finish speaking.

Sweat rolled down Fran's face. *What the hell was missing from that drawer that was so important?*

In a not-so-distant area, deep below the ground from CERN and the HEPA lab, the director sat motionless in his chair in a large office; the holographic data, photos, and video moved steadily across the room.

In the dark room, he leaned on the arm of the chair and pulled and twisted and stroked the long, thick, greying goatee on his chin. There was an alert on the holographic data in front of him: with the flick of his hand, the image of another man came into focus.

"Yes?" was the director's slow and deliberate question to the man.

The other man spoke nervously. "Director, I got a message from one of my sources that CERN and the HEPA lab are in lockdown, and there are possibly two deaths. What is going on? What is going to be your move on this? I need to alert the President."

"You will do no such thing. This information stays between you and me until I know more.

Understood?" The director was steadfast.

"I am the liaison to your program; I don't take orders from you." The man's voice had changed from nervousness to irritation.

"And I suppose the President would also like you to leak all the financials you approved for this program? If we are suddenly being honest and transparent with everyone, that is." The director's deep voice filled the area around him.

"I will not take the fall for ......" The director cut off the man, raised his voice, and spoke. "You will do as I tell you to do and say what I tell you to say, Mr. Secretary of Defense. If you were to leak the detailed financials for this program, not only would the HERB Mission be exposed worldwide, but the unconscionable amount of money being spent here by your government would be exposed as well. I will make sure the leak gets traced back to you, and I'm confident it won't be long before I meet your replacement."

The screen where the man had listened to the director's threat had gone black with a swipe of his fingers. The room became silent, and the director continued to pour over petabytes of information streamed to the holographic projector.

Above the director's office, through the 200 meters of sand and rock and soil, the nighttime Swiss public moved about, oblivious to the conflict raging below the ground that would determine the world's fate.

Chapter 20: Catching a Shadow (23.11.1981)

Doctor Leonardo Hoffman and Bruce Hayden sat in the basement at the ramshackle house at the end of Fremont Street, brainstorming their next move. Leo sat silently; his left arm was in a sling and had a white piece of gauze covering his left eye, which was still badly swollen. Bruce and Leonardo stared at the holographic projections that filled the room around them: flow charts, computer analysis, and details of the ongoing scan for any communication with the other HERB device. They didn't speak for some time. It was Leonardo who broke the silence.

"What would I do if I had just been unexpectedly transported to another time after trying to sabotage the HERB mission? I had knowledge of the mission and some knowledge of the existence of mission passcodes; I was smart enough to get an undetectable firearm into a highly secure facility and ambush the leader of that mission group."

Leonardo thought aloud as both men stared at the holographic light. "I had enough knowledge to look for the HERB device after transport, and I knew how to power it down." Leonardo gasped and covered his mouth with his good hand.

He thought for a moment. "I remember seeing the HERB device in the sand, and I must have blacked out before I got to it, or he moved me to get to the device. Did the HERB device malfunction and shut down after transporting two people? It's not supposed to do that!" Leo's voice broke as he spoke. Bruce interjected into the thought process.

"No, the device was functioning when you arrived, Leo; remember what I told you?"

"Yes, yes," muttered Leonardo, still in his train of thought. "He knew enough to power down the HERB device with my biometric signature and how to do it."

Leonardo looked at Bruce with panic in his eyes. "I can count on one hand the number of people who know that much about the device, Bruce. You, me, Fran, Mitch, Gus...." he counted with his fingers up in the air before he looked at the floor. "Oh God, Gus is dead...." He brushed his hand over his face as he exhaled, and his hand recoiled after accidentally touching his painfully swollen eye. He took a long, deep breath and exhaled it, and he could feel the tears welling up mostly in his swollen eye.

His breath shook as he exhaled; he was stressed.

"I don't know who this guy is, but he knows ultra-top-secret information about the HERB device and is out to sabotage this mission. What would you do?"

Bruce thought for a moment. "Well, I would find shelter first and food, and perhaps he was injured from the time travel, so maybe he would find a hospital?"

"No, no," Leonardo muttered as he shook his head. "He wouldn't do that because he would risk people asking questions. If he had enough strength to leave that area on his own and had the wits to grab the HERB device and deactivate it, he would not be badly injured, if he is injured at all. Plus, there was no blood in the area other than mine, right?"

"No," Bruce added, "and there was no sign of an awkward or unusual gait from the footprints I captured with the nanocam from that desert area."

"Right, right, and AI analyzed those images, right?" Leo asked.

"Yes." Bruce had lifted his head from his hands long enough to answer, then put his head back down on his hands.

Bruce and Leonardo were both lost deep in thought.

Leonardo broke the silence the second time, "He was masked, he had a voice alterator, he had a well thought out firearm. He had that powder with him that he put on my arm and face after he shot me. It took away a lot of the pain, some kind of anesthetic powder. It's something synthetic, similar to cocaine, I'm guessing. This guy is smart and in a scary way." Leo's voice was getting louder as the memory pulled him into the incident.

"He's like a shadow because we don't know anything about him, where he came from, where he went, or how to find him."

"Except for his shoe size and vague foot treat pattern," Bruce interjected.

"Yes, but we also have the rudimentary chemical analysis from the AI unit." The AI sprung into action, hearing the conversation and projecting the chemical analysis of the powder residue in the air space in front of the two men.

"That white powder was cocaine," mumbled Leonardo as he looked at the information,

"That's primitive medicine; where would he even get powdered cocaine? Not even synthetic cocaine? Hmm."

Bruce continued his train of thought as if Leonardo had not spoken. "If I were him, I would be ditching that footwear first. So, we know his shoe size. Men's size 9, regular width. *How average.* That will give lots of false hits if we set up AI to monitor footwear sizes in town."

Bruce stood up and stretched. "Leo, let's take a break; you need some rest. I have to be back at the university with Doctor Nault tomorrow, and you must be searching for, for," he stopped. "For Shadow. That's what we should call him for now. It might make things easier to give him a name."

"Shadow it is," said Leonardo. "Shadow, prepare to

be hunted."

Hours later, while Bruce slept, doctor Leonardo Hoffman sat in the basement fortress of technology. *He was the reason they were in this mess; he needed to make this right. This mission could not fail. Shadow could compromise the fate of humankind should he be able to infiltrate this mission.*

Leonardo thought as he poured over reams of data: area maps, topographic maps, transportation routes, motels, car rentals, and weather data. *We must assume he knows the mission plans, locations, players, and organizations involved with Bruce's mission.*

"How do I find this guy? Think, Leonardo, think." He stared at the information before him, gently rubbing his fingers on his sore temple.

"The electronic infrastructure here is non-existent; there is no signal I can grab unless I set up a device to tap into closed circuit security footage. No, that would be too cumbersome. I'd have to have a unit directly in contact since there is no wireless signal in this time. I'd only be able to capture television or radio signals, which would be mostly useless. The overwhelming majority of video security is hard-wired and records to those giant disk things - those VHS tapes." Doctor Hoffman was deep in thought.

"With only the five nanocams that Bruce has, and one already deployed at Kevin Nault's office, I will have to run an optimization on what configuration is best to cover the most area around this house and the university," Leo thought aloud, mainly for the benefit of AI. Doctor Hoffman entered some commands into the holographic screen into the air with his finger movements.

After a few minutes of programming in the code conditions for optimization, along with AI's intuitive knowledge, the holographic screen displayed a map of the area with the five nanocam locations flashing in red. One in Kevin Nault's office, and the other four spread precisely between the university and the house on Freemont St. There was also a countdown time indicator for the individual cameras indicating how long each would likely be in the proposed location before relocating it as the mission progressed.

As he worked through the night, he began another series of code entries into AI to calculate the precise wavelength of the quantum energy signature from the three different time travelers. Meanwhile, AI quietly began its petaflops of calculations, scanning Doctor Hoffman to determine the exact energy wavelength or signature that would linger the longest.

"I wonder if Shadow thought he would be traveling back so soon, and did he equip himself with what he would need? Does he have the appropriate money?" Leo continued as AI listened. The holographic projector interrupted the doctor's thought and displayed the optimized frequency of quantum energy that they would be looking for. He examined the information before him and thought for some time.

The signature AI had calculated that they should be searching for was in the ballpark that he had speculated; however, the total dissipation time of the quantum energy projected by AI was approximately sixty days. Leonardo checked the numbers closely as they cast in front of him and guessed that the nanocams could be programmed to pick up that energy signature for just over half of that time.

"Twenty-eight days after today," he mused. He sat there, staring, calculating. "How do I increase that detection time? How do I modify the instruments to pick up the signature for a longer time? I'll have to scan Bruce in the morning with the new algorithm, test the nano cams, and verify how long we have to find Shadow before the signal goes dead."

His mind was still wandering as he shut down the holographic projector with a wave of his right hand. He rubbed his aching left shoulder as he got up and lay down on the makeshift bed in the basement, which consisted of the stretcher from the desert rescue and a thin wool blanket.

It's been a crazy 36 hours. I wonder what is happening at the lab. I wonder if Fran and Mitch are ok. The doctor dozed off to sleep, with thoughts of a different time weaving through his head.

Chapter 21: The Unexpected Time Traveler (23.11.1981 22:35 hrs.)

Shadow sat on an old stained blanket and stared at the entire collection of possessions laid out on the derelict single bed: the deactivated HERB device, the polycarbonate anti-projectile vest that contained a multitude of small bundles of 1980s American bills, reams of tightly folded note paper, a custom polycarbonate magnetic powered handgun with eleven ceramic bullets, a mask with voice alterator, and several different fake photo identifications.

The 22nd-century watch on Shadow's wrist was similar to Bruce's; it was far superior to any supercomputer from this time; it stored reams of data in its memory: locations, dates, names, and detailed mission plans. Removing the ballistic panels from the vest pouches and rearranging the Velcro straps made a pretty good packsack in which Shadow loaded up all the possessions, save for the paper map of the area and a couple of neatly folded $20 bills.

Shadow disassembled the handgun into several pieces, with the ceramic rounds hidden inside the magazine compartment, and placed them into the bottom of the packsack. Time was of the essence, and even though Shadow's time travel was at an unexpected juncture in the plan, it was still a fortunate occurrence that needed to be taken advantage of. Shadow still ended up in the right time frame, with most of the tools to get the job done. Packing up all the items from the bed, Shadow took one last glance at the old abandoned motel room and maneuvered through the smashed, leaning door and into the dark, drizzly night.

The stranger walked silently in the night, organizing thoughts and plans. Bruce Hayden, who also didn't belong in 1981, would likely be hunting, watching, and planning.

Had I not been in such shock after being transported, I should have finished off Doctor Hoffman, just to be sure, Shadow thought. *He was in bad shape; hopefully, he didn't last long in that heat. Besides, I would have wasted another round on him.* It was uncharacteristic of Shadow to leave a loose end like that. *I'm just glad I found the time-travel device so quickly.*

Images flashed through Shadow's mind: people fought, so many battles, so many people killed—so much blood. Shadow's first memories of 1981 were disjointed, stumbling around the desert sands; the images were clouded and fragmented memories of the last time Shadow saw Doctor Hoffman. Shadow was desperate for clarity: had the last bullet fired in Doctor Hoffman's office hit the doctor in the face? Shadow's mind only showed an unconscious doctor with blood on his face and neck.

Shadow's head shook with the confusion of the memories that came flooding in, overlapping sights of many dead and injured people over the years. Time travel's effect on Shadow's mind was not wholly unexpected; it was an unknown that Shadow accepted before the attack on the HEPA lab.

Several minutes later, with head straightened and mind more focused, Shadow put a positive spin on the multitude of disjointed memories. *No matter. Even if Doctor Hoffman is alive here, then I will eliminate him.*

Weaving through the dimly lit city streets, expertly

carving the path of most darkness, Shadow started the long journey to the first waypoint in the mission.

Chapter 22: Act Normal; Whatever That is (24.11.1981 08:00)

The morning light shone sweetly into Bruce's bedroom window; however, Bruce had been working in the basement for hours now.

Bruce and Doctor Hoffman sipped their morning coffee slowly, absorbed in thought as they examined AI's reports on nanocam locations, probable locations of Shadow, and algorithms to calculate and detect the quantum energy signature of post-time travelers.

AI had scanned Bruce immediately as he walked down the basement stairs two hours earlier, and he found Doctor Hoffman busy at work writing some hand calculations on a scrap piece of paper.

"Whoa! Good morning to you, too!" Bruce joked as AI silently began scanning him before he even finished coming down the stairs into the basement work area. Bruce's eyes glimmered as the scan reached his face, reflecting and refracting the subtle blue light. The light dimmed, and Bruce continued to his chair and sat beside Doctor Hoffman.

"Morning, Leo." Bruce thought for a moment. "Maybe AI can develop an algorithm to make this coffee taste better." Bruce chuckled at his joke as he looked down at his sad cup of coffee. There was no response from Doctor Hoffman. Bruce looked at the doctor, "What's with the morning scan? Still checking if I'm a redhead?" He chuckled again at his comedic prowess, but less so this time, as his tough audience gave no reaction whatsoever.

"Oh, ha-ha, yes, the coffee could be better," Leonardo mumbled as he looked up briefly in Bruce's direction, only half engaging with the conversation as he returned to his pencil scribbles on the paper in his hand.

Both men stopped and watched, anticipating AI's analysis, as the holographic projector gave the not-so-good news.

"Worse than I suspected," Leonardo muttered as both men read the information.

"What?" Bruce asked as he tried his best to keep up

with the doctor, "Looks like we can scan for Shadow; that's good."

Doctor Hoffman stood up and started pacing around the room.

Bruce watched him pace back and forth as the doctor ran his hand through his hair, with a noticeable flinch over his still-aching temple.

"Ok, so here it is." Leonardo stopped to gather his words. "You and I both are emitting a specific energy signature since we time traveled, but, over time, that energy dissipates down to zero."

"Yes, got it." Bruce was patiently following the intense doctor.

Doctor Hoffman pointed to the curve on the projector, which showed an initially slow decay and then a rapid decrease to zero. Bruce continued, "It looks like it gets to zero in around 28 days."

"Yes, but look at the type of energy we are dealing with; Alpha radiation is what we can detect the best." Leo's voice quieted, "Alpha radiation destroys human cells the most readily, so when I designed the HERB device, I had to allow alpha radiation dispersal away from the body so that there weren't tissue burns due to time travel. There is a brief burst of Gamma radiation when the Einstein Rosen Bridge opens and another when it is closed. The energy transfer to the body when the body cells transport across the bridge is negligible, but only because I designed the energy to fall away from the body as they decay, just like dead skin cells slough away from the body as they die.

The gamma radiation from the bridge being opened and closed is long gone, although I could most likely detect the signature at the location we were transported to in the desert. Other than that, the gamma radiation will be undetectable. This morning's scan shows zero gamma signature from your body, which was the same for me, which I scanned last night.

"Hmm." Bruce took in all the information as the doctor spoke. He sipped his coffee and looked into his mug disappointedly; at *least it had caffeine.*

Leonardo continued, "The next most penetrating radiation is Beta radiation, which gets trapped by aluminum, other metals, or dense materials. As you can see, the beta radiation is virtually gone from your body, Bruce, but it is still detectable in me. The projection changed graph images as the doctor spoke, as AI followed his logical progression. The amounts of beta radiation will be detectable from Shadow or me for about six more days, but it has to be a line of sight to detect it. That means the nanocams will only pick up Shadow's Beta radiation if he physically goes past the field of view of the nanocam.

So, we come back to the Alpha radiation."

The holographic graph again changed to the original chart Bruce looked at as Leonardo continued, "The Alpha radiation lasts the longest because I designed the energy to decay away from the body and not penetrate human tissue. The problem with Alpha radiation is that..."

Bruce finally interrupted the doctor, "That Alpha radiation can be stopped by something as thin as..." He grabbed Leonardo's paper calculation from the chair where he sat. "As thin as a piece of paper."

"Correct." Leonardo came to sit down and snatched his piece of paper from Bruce on the way to his seat.

"So, direct line of sight, we have about seven days for Beta radiation and twenty-eight days for Alpha radiation."

"Doesn't it absorb into clothes?" Bruce thought aloud.

"Yes," Leo replied, "the clothes Shadow wore when he was transported will show some Beta and Alpha radiation, but not any more than his body itself. If he's changed clothes, which I assume he has, we can only detect the radiation from his body and the weaker radiation that has soaked into his fresh clothes.

Bruce jumped in with his usual positivity, "So, upload the nanocam map to my watch, and I'll install them right now. We don't have a moment to lose."

"Already uploaded to your watch, Bruce, and the nanocams are in the container, there on the table," the doctor continued with a wave of his hand and pointed finger, "and don't forget, Shadow knows what we look like, but we don't know what he looks like, so please be careful out there."

"Gotcha," Bruce said as he tied up his packsack and flipped his apple into the air, "Do what I do, and keep an eye out for Shadow...whatever he looks like."

"Well, Bruce, I should tell you too; I've been using AI to visualize Shadow's face. From my interaction with him at the lab, I have already approximated that he is six feet tall. Also, he has brown eyes. That is unless he was altering the color of his eyes.

So, approximately your height, a little thicker than your build, and the rest, I'll try to remember more.

"Got it." Bruce now had one foot on the bottom stair. "I'm going to be late for class, Leo. I'm really not that worried about it. It's just my new reality, and I've got it. We've got it. I'll act normal - whatever that is. Later!"

Bruce leaped up the stairs, out the door, and settled into his trotting cadence on the way to class. He glanced at his watch as he rounded the first corner. Bruce took a deep breath once he was out of sight of the house. Contrary to what he was telling Leonardo, he worried about Shadow being out there somewhere and had to keep his guard wherever he went. He also didn't want to stress Leo out; Leo was a huge asset in planning and hunting Shadow, but not worrying about Bruce was not Leo's forte. Having Bruce away on a mission was one thing, but he knew that Leonardo being on the mission with him was another ballgame entirely.

Bruce would have to bear the brunt of the physical work to hunt Shadow because Leonardo Hoffman, as brilliant of a scientist as he was, was not a physical person. His recent injury, memory loss, and dizzy spells made the circumstances more obvious that Bruce would have to do all the leg work.

I wish I had some extra blood nanobots. That would take care of a lot of Leo's problems, Bruce thought.

At least I do have Leo here with me. That is a distinct advantage, even with the injury. Leo is a formidable opponent by himself, but we will get this Shadow character together.

Bruce got to the school doors and ran up the musty, air-filled staircase to his 9 a.m. physics class with Doctor Yu. Nobody noticed that as Bruce walked into the hallway near his classroom and stretched his long, lanky arms, he placed a nanocam on the side wall of the hallway outside his classroom on the third floor.

Leonardo, who hadn't moved from his seat in the basement on Fremont Street, now had fresh eyes to look out for Bruce while in class.

A simple vibration on Bruce's watch would alert him to any suspicious activity outside his classroom. In the background, AI was scanning, watching, waiting.

Chapter 23: Revising the Mission (25.11.1981 00:05 hrs.)

Shadow sat in the small, dank basement apartment in Tucson. Fortunately, this part of the city was known for its high drug and crime rates; making it the best place for Shadow to hide out when not on the move. The area's name was Iron Horse due to its proximity to Iron Horse Park. Shadow thought the name was appropriate for the mission; Shadow was the Iron Horse that would ride in and stop the madness of manipulating history with time travel.

The one hundred and thirty-eight kilometers from Willcox to Tucson was quite a distance to travel on foot, even for Shadow. So, Shadow had hopped a couple of greyhound busses along the route when possible and walked the rest of the distance after appearing in 1981, four days prior.

Shadow walked down the narrow flight of stairs to the one available room, the one that the building manager referred to as "that little fire trap in the basement." This made negotiating the rental of the space easy.

Shadow paid in cash up front for the living space and, within a day, had installed new locking hardware on the doorway. The now secure door was more for the safety of any person unlucky enough to try to break into Shadow's space than for Shadow's protection.

Shadow didn't want any attention from the authorities, and laying low in the drug flop house was the way to achieve this since police generally stayed away from that area of the city unless they had to be there.

The room was small and had been used as a shooting-up room recently by the looks of the paraphernalia and stains on the floor: burned spoons, used needles, a tourniquet, and a few minor blood stains.

The only items left in the room after Shadow threw all the unnecessary things into a dark corner of the basement hallway were the bed, a glorified cot, and a lamp with a light bulb that surprisingly worked. Shadow was used to sleeping in much more extreme conditions on a mission; any bed not in the jungle or tundra was a bonus.

Sitting by the dim light of the lamp, augmented by the street light peeking through the grated window, Shadow worked on refining the mission parameters. The time travel destination date was a couple of months later than initially planned; however, as always, contingency plans had been made for multiple alternate scenarios. Shadow folded the bus schedule pamphlet after skimming the times and laid down, committing the bus schedule to memory as sleep descended.

Chapter 24: Such a Simpler Time – A Brief History of North America Part I

1981 was a time in American history that was far simpler than 2131. 1981 saw the release of basic personal computers such as the Commodore 64 and IBM's personal computer 'PC.' These computers could only perform essential functions and run some rudimentary games.

There were no luxuries like Bluetooth, Wi-Fi, streaming, texting, or cell phones, which would be readily available to the public within the next 25-30 years. However, with the advent of these technologies in the late 20th century, they magnified and compounded systemic issues in a country rife with racism, corruption, sexism, and homophobia.

Over time, as some of the American public evolved to try to become more inclusive, more accepting, and less discriminatory, the early 21st century gave rise to race riots and the LGBTQ2+ movement. In that time, for all the social progress in some parts of the country and some parts of the world, there were pockets of the United States of America that resisted social change, including white supremacy groups, government lobbying groups, anti-abortion activists, and freshly voiced anti-vaxers.

Then, in 2006, the first central social media platform was released to the public. Its intended function was to bring people together to communicate and share memories using internet technology; however, by the 2020s, its users, disruptive agents, and countries had warped the platform so much that it helped spawn the information wars over all social media platforms.

The information wars plummeted America into a civil war that spanned almost three decades. The problem was two-fold: social media contained advanced algorithms (at the time) that not only provided a drug-like addiction to its use but, by delivering topics that the user was interested in, created large platforms for people to spread misinformation.

In the not-so-distant past, these people may have had dozens or hundreds of followers at most, but social media allowed anyone to say anything to a large audience, regardless of its truth. Future societies agreed that the spawning moment of the information wars was the 2016 campaign leading up to the election of President Donald J. Trump, which then was followed by scandals numbering in the dozens, which included the worldwide COVID-19 pandemic, and the storming of Capitol Hill in Washington DC. All these events, fueled by social media misinformation, allowed information to be warped and disfigured by prominent figures like Trump.

The dawn of the misinformation wars allowed people with no previous background or knowledge to research any topic on a social media platform and lead themselves to believe that they knew better than the government health professionals or the rest of society. Starting with the COVID-19 pandemic in late 2019 and lasting over three years, millions of people died worldwide; many died because vaccines were not available to poorer countries, but hundreds of thousands of people died worldwide because they had been victims of misinformation and chose not to be vaccinated.

The anti-vax movement, which had started in the early 2000s, came to a fevered pitch magnified by the global pandemic. By the late 2020s, the group was flagged by governments as having violent extremist members. The 1960s hippie movement that was generally anti-government was comparatively peaceful; however, the second generation of hippies, dubbed covidiots, was a much more violent, extremist generation of anti-government people.

Fueled by misinformation and swollen egos, these new extremists stormed Capitol Hill in 2021, carried out bombings of vaccine suppliers and their supply chains in 2025, and began overthrowing state governments by late 2027. By then, the misinformation war was in full swing and lasted another decade before America fell into a decades-long civil war called the Great Divide.

The Great Divide was the second great civil war in America's history that descended like a fierce storm cloud on an already exhausted nation. Race, religion, sexuality, abortion, vaccinations, police authority, and government powers were the basis of fighting between the three main groups.

The Government: would be gradually dissolved, broken down, and eventually transformed into a better representation of the people of America.

Trumpists (who called themselves the New World Trump (NWT)): people who followed the words and ethics of former President Donald Trump. The Trumpists would continue campaigning for their new world years after their leader, Donald Trump, died in 2030. To the other groups, the Trumpists were known as 'the Anarchists.'

The CSA (Common Sense America): people who were tired of misinformation, tired of violence, tired of racism, homophobia, transphobia, and the old system of Government. All three groups contained extremist-inclined people who wanted to use violence to further their cause, but after several violent skirmishes between the three different factions, all had become violent to some degree as a matter of self-preservation.

What remained of the American military initially protected the Government; however, more than half of its service people left to support one of the other two warring groups. The conflict ripped the country into three distinct parts, each controlled by one major group of people, and it left a sizeable portion of the country, which included a strip of land approximately one million square kilometers running north-south down the middle of the country. Thus, the Great Divide was well underway.

This strip of land was called 'the grey zone' for several reasons: the majority of the swath of land had no organized electrical grid, there was a barely functioning food supply chain, and living there was a group of people who had abandoned any affiliation in the civil war. The grey zone people had formed several communities in the more remote areas of the grey zone to minimize their involvement in the fighting.

These people, who called themselves the 'middle people,' lived mainly off the land. They hunted and foraged for food and embraced living in nature. They lived the middle way, which was primarily peaceful and sustainable.

To the rest of the people who were at war with each other, they were known as greys. Climate change was a major driving factor in the Great Divide, as large groups of people flocked to coastal areas with more stable infrastructure, such as electricity and clean drinking water. Global warming raised the country's average temperature by several degrees, and the populated areas with poor infrastructure in the grey zone became baked and dry with the heat. Underground aquifers had all dried up, and millions of acres of land were abandoned in the search for fresh water.

Due to the civil unrest and subsequent disorganization, the United States's fuel supply was badly interrupted. The opposing groups fought over and seized fossil fuel-producing areas, while Canada cut off its neighbor from its supply of fossil fuels due to the unrest south of its border.

America was so disjointed and disorderly as she focused all her remaining strength on fighting herself that it didn't take many years before she had metaphorically and physically burned herself to the ground. It would take decades for her to rise from the ashes. Re-born.

Meanwhile, in 1981, in the United States of America, where there was a secret battle to protect the Earth, the lack of technology worked mainly in favor of Bruce and Leo. Messages were sent by mail, which a person delivered, and if someone wanted to sabotage a mission, there was little technology to track people other than do it the old-fashioned way: watch and follow and make a plan to disrupt that person's task.

Bruce, Leo, and Shadow's 1981 possession of 22nd-century technology was a distinct advantage as they jockeyed for a front-row seat to the discovery of the millennium.

Chapter 25: The Invitation to the Telescope (28.11.1981)

"Nicely done, Bruce," Doctor Nault complimented as Bruce watched him read over a proposal to make their Saturday morning theories into a more concrete, programmable reality. "It is remarkable how you came to me and how we have gotten here," Kevin's voice trailed off at the end of his sentence; he was absorbed in thought.

Bruce interjected. "Yup, it's cool how this has happened, right?" He looked to Kevin for an answer, but he was still thinking.

"Ahem.... doctor Nault?"

"Oh, yes, Bruce," Kevin stammered as he looked up at Bruce.

"I thought that maybe I could start coming with you to the observatory on Tuesdays and Saturdays. I could work on the programming code for this experiment. I won't bother you in any way, I promise. That environment will help me focus strongly on what we want to do with this new theory." *That was an understatement.*

Bruce still had time, but he needed to broach the telescope time with Kevin so that the mission could progress. He did have another ulterior motive now; he could not take a chance to leave Doctor Nault alone at the observatory with Shadow around.

Doctor Nault thought for a moment. Bruce waited with bated breath, knowing that if Kevin said no, he would have to plan a trip to put the nano cams up in and near the observatory. That would cost Bruce a lot of physical work and time.

Kevin looked at Bruce. "Sure, my boy, why don't you join me starting Tuesday this week? That way, we will expand our Saturday mornings and move forward with our work. My telescope time tonight was rescheduled, so I won't be there until Tuesday."

Bruce exhaled without even realizing he was holding his breath. "Great! I can bring a student computer with me on Tuesday, and I can start writing some code with the goal of allowing the computer to accept raw data from the telescope."

"No need, Bruce," Kevin put up his hand. "I have a very nice little setup at the observatory, and I leave my computer in one of the storage rooms. You can use that one while I use the computer that controls the telescope. I'll bring some snacks; it'll be a long night. I hope you can keep up with the lack of sleep!"

Bruce laughed, "I'll give it my best, professor." *I could outlast you any time in the sleep deprivation department, buddy,* Bruce thought as he left the meeting and began to walk home.

Bruce began planning how to keep all their collaborative work encrypted on Doctor Nault's computer at the observatory so that Shadow could not steal the information. He could easily interface with the archaic computer at the observatory with his watch and store their information covertly in his watch on his wrist.

AI had been running simulation after simulation back at the house to predict Shadow's first target. The telescope would be an obvious target for Shadow, and Leonardo was already working on manufacturing a sabbatical-type leave for Doctor Nault to work at a different telescope. They needed a reason to make it realistic for

Doctor Nault and Bruce to move locations temporarily until they made their significant discovery of NH-1983.

That discovery was almost 18 months away, so they would have to move locations as soon as possible to remove the distinct possibility of an attack on the Steward observatory. They needed to remove the observatory from their chess game with Shadow. The Steward observatory would be like the Queen in a chess match; she was a telescope they had depended on almost exclusively to discover NH-1983. Secrecy was paramount in the modified mission, and Leo was working on all the loose ends to ensure that nobody knew precisely what telescope Bruce and Kevin would be working at. It would be a race to make the telescope switch while not allowing the mission to be compromised via the Steward telescope.

Until they made the observatory switch, Bruce and Leo had to be on high alert. An attack by Shadow could be devastating, and they nicknamed the part of the mission dedicated to taking the Steward Observatory out of play for Shadow, "Make to the Queen!" Bruce, Leo, and AI had postulated that Shadow was a highly trained professional espionage operative, and from Leo's interaction with

Shadow, also a military-trained operative.

AI gave the probability of Shadow expecting that he would be transported to 1983 when he was in Doctor Leo's office to be .03 or 3%. However, the likelihood of Shadow knowing that he would be sent to the exact location in the desert, plus or minus a few months of Bruce, in the event of an uncontrolled time travel is 0.83 or 83%. This was assuming the breach of information that allowed Shadow to know about the mission codes, parameters, and the location of the offices of Doctor Gus and Doctor Leo.

The probability that Shadow would have planned that he and Doctor Leo would be transported simultaneously to the desert was rounded to 0%. Not even Doctor Leonardo Hoffman himself could have known that two people could have time-traveled at the same time when he attempted to escape from his office with the gunman standing over him.

AI followed through with the logical progression and postulated that not only was Shadow likely unexpectedly sent to 1981 from Doctor Leo's office, but that it was highly probable that Shadow knew how to manipulate the date and time settings on the HERB device.

He had planned to use Leo's fingerprints and facial scans to allow him to go to his desired date in time in the past. The probability AI calculated for those events to be actual was 0.94 or 94%.

Bruce completed his walk home in thought and threw his ball cap at the back of the kitchen chair as he entered the house. He missed the chair as he sailed downstairs and met with Leonardo, who was sitting quietly watching the AI holographics. Bruce sat down, and the two men watched as the AI continued to display probability graphs and data until, with a wave of his hand, Leo shut down the holographic projector and looked over at Bruce while he leaned back in his chair.

"We have to proceed with the start at the Steward observatory on Tuesday, Bruce. This being Saturday, we have today, Sunday, and Monday to scout out the observatory and ensure it's not been tampered with before you drive up there with Kevin. I heard Kevin got the memo that his telescope time had to be canceled tonight."

"Yes, smarty pants, he got the memo," Bruce smiled at Leo's orchestration of the cancellation.

"I'm good for scouting the Steward telescope this weekend," Bruce said in a chipper voice, "there's no way Kevin and I are showing up there Tuesday evening without doing a rekey ahead of time.

"It'll be a bugger of a hike, but at least I have three full days to do it. I am okay missing my class on Monday morning."

"A hundred and thirty-eight kilometers, Bruce!" doctor Hoffman yelled indignantly. There's no way you could hike that." Bruce cut off Leo.

"I was joking, Leo. I would have to steal a car or hitchhike for at least part of that distance. It would be a risk but a calculated one. I wouldn't get caught."

"Well, I've been thinking about that," Leonardo said, still with his brow furrowed at the idea of stealing a car, "what if I rent or borrow a car, and we drive up together? It will save you a ton of physical work and allow me to get out of this place for some fresh air."

"Not a rental company." Bruce interjected. "it's too risky, and don't forget, Shadow knows what we both look like, so why would we put two faces out there instead of one? Besides, I need you here to work on make to the Queen."

Leo thought for a moment and examined his watch before speaking. "Listen, Champ. I have been cooped up in this house for over a week and haven't stepped outside even once. You're trained for this stuff, but I'm just a scientist. I don't have the same mental fortitude that you do. I need fresh air; I need to get out of here for my mental well-being, Bruce."

Bruce pondered for a moment. Leonardo had been calling Bruce "Champ" since he was a boy, and it gave Bruce a sense of security as he thought of working with Leonardo in the lab when Bruce did not have a care in the world—relatively speaking.

Bruce looked at Leo. "We can go together, but only if we can borrow a car. And no rentals. Shadow probably has eyes on the rental places."

"Done!" said Leo, "Road Trip!"

"Pfft!" Bruce mocked at the joke.

"So, does Miss Rosie have access to a car?" Leo asked.

Chapter 26: Make to the Queen!

Miss Rosie's face was red when she came to the door, and Bruce braced for a tongue lashing he had endured many times, although he couldn't think of why she would be mad this time.

She flung open the door and started yelling with her eyes still closed. "You think you can just come over here...." She stopped when she saw Bruce standing there. "Oh, it's you, Bruce. Sorry, I've been shooing off my deadbeat nephew, who keeps calling and asking for money. I keep tellin' him, I ain't got any."

Bruce segued into his well-rehearsed lines, "Well, Miss Rosie, I don't know about your nephew, who I'm sure is a good lad." He stopped to wait for her reaction. She sheepishly nodded as she checked her curlers; they hadn't moved out of place.

"He is a good sort, you know? But I just can't be giving money to every Tom, Dick, and Harry that asks me."

Bruce went in for the kill. "I'm sorry that you don't have any money, Miss Rosie, but perhaps we can come to an understanding where I can help you out. Financially speaking."

Miss Rosie's eyes got that familiar greedy look in them. She was a good person, but money was just irresistible to her.

Bruce continued, "Um, Miss Rosie, I was wondering if you knew anybody around who has a car I could borrow. I would pay a rental fee to use it for the day. You see, I want to go to the State Library in Tucson to learn about my relatives. Now that I have a part-time job at the university, I have some money to do that kind of thing, you know?"

"I don't have a car," Rosie seemed deflated. She looked down at her poorly manicured toenails.

"I know you don't, Miss Rosie, but who would you ask if you needed a car to borrow for a day or two?"

"Well, I am not one for drivin' Bruce; you already know that much about me. But if I did need a car, I'd ask Jimmy Neil, who lives around the corner. He's got that big car that he hardly uses and money problems with his wife, leavin', and all. Talk about money suckin'; he should ditch that car to save the money on gas! I suppose it would just free him up to fill his gob with more booze." She pointed her finger at Bruce, "and that doesn't get repeated, Mr. Bruce."

"Of course, Miss Rosie, I wouldn't tell anyone. It's not my business." Miss Rosie nodded in reply.

Bruce had to put this next part delicately so as not to offend Miss Rosie's good sensibilities. It's a good thing that they were more like reasonably well-intentioned sensibilities.

"Um, Miss Rosie." Bruce stammered and shuffled his feet as he spoke. "Suppose you asked Mr. Neil if you could borrow his car and let me use it so I can go to the State library but keep the last part between us?"

Bruce looked at the ground before adding the slam dunk to the equation. "Um, I'll give you the same amount that Mr. Neil charges for it to be our little secret. I get nervous around Mr. Neil, what with his drinkin' ma'am. My dad drank like that," Bruce lied.

Miss Rosie looked at Bruce's forehead as he stared at the ground. She reached out and put her hand on Bruce's shoulder, "I understand, you poor boy. Well, I don't like the idea of lying, but," she thought some more, "I also know what you're talkin' about with the alcohol, and you just want to go to the library for the day, right son?"

"Yes, ma'am," was Bruce's quick reply. "I might end up driving back the next day if I get somewhere looking for my relatives, ma'am."

"Well, for that kind of responsibility, Mr. Bruce, you need to give me some insurance that you'll return the car as you found it. After all, my good name will be borrowing the car."

"Yes, ma'am. I can give you twice what Mr. Neil charges." *You sly woman,* Bruce thought as he tried to fight off a smirk.

"Now, that's better. I'll be talkin' to Mr. Neil this afternoon."

She waved her arms as if to shoo Bruce off her porch, partly with guilt for taking advantage of the situation.

Bruce hopped off the porch. "Yes, please, ma'am; I'd need to leave tomorrow morning to get to the library bright and early."

Bruce awoke early the next day in case Miss Rosie failed to come through with the car they needed to get to the observatory. He peered out his bedroom window, and with a whoop, he ran down the stairs, yelling, "Leo! Get up! Get your stuff; we've got a car!"

Bruce opened the door and saw the old, grey, beat-up '76 Chevrolet Impala sitting in front of Miss Rosie's house. He could see a note on the driver's side window, so Bruce ran out barefoot and snatched the letter from the window.

Mr. Bruce, Mr. Neil was as drunk as a skunk and on quite the bender. He wanted to give me the car for free, but I insisted I pay him $10 on my return. With a full gas tank, mind you. Keys are in the gas flap. – Miss R.

Bruce ran back to the hatch that covered the gas tank receptacle, grabbed the keys, and boomed inside the house to find Leo coming up the stairs with a bag in his hand. "Computer, nanocams, first aid kit, money," he listed out loud as he looked at the bag.

Bruce closed the front door before anyone could see

Leo, adding the $30 to their needed money. Bruce spun the large car key ring around his index finger proudly.

"I'll just put the money under her front door mat like I've done before," Bruce said as he stuffed his mouth full of the apple he had just grabbed from the refrigerator. He rummaged around in the fridge and gathered some food to bring for the trip, the key ring still dangling on his finger.

Leo stood there with the bag in his hand as he closed the refrigerator door. "I've set up the computer so that AI can scan out the car's front window as we drive, especially when we get closer to the observatory. I don't want to be ambushed by that trigger-happy moron today."

Leo continued somberly, "Bruce, if things should go badly at any point today, I want you to take the equipment and the car and leave immediately. Don't wait for or attempt to save me if I'm injured or captured." Leo pointed to the injury on his face, which still looked raw.

Bruce stopped chewing his apple and looked at Leonardo. "Leo, it'll be fine. Don't be so dramatic. Besides, I doubt he will expect us to roll up in that big boat of a car in the daytime." Bruce smiled.

"I'm just saying," Leo put his hand on Bruce's shoulder. "I'm reminding you that you are the most important part of this mission, and you must be successful at all costs, especially now that we have a saboteur in our midst."

Bruce knew what was at stake, but he didn't want to stress out Leo more than he already was.

"Leo, we will be fine, I promise." Bruce touched Leo's shoulder and repeated, "I promise."

Leonardo nodded silently and returned to the lab to ensure they had everything they needed.

Five minutes later, the grey Impala stopped around the corner out of sight of both Miss Rosie and Mr. Neil, and a figure awkwardly ran out of the bushes at the side of the road, got into the passenger seat of the car, and slunk down in the chair as the vehicle rolled away. Leo slid the laptop AI unit onto the long, massive dash of the car under the windshield, and it immediately began scanning everything around it. AI would continue to monitor the entire one hundred and thirty-eight kilometers of the drive to the observatory and back.

As Bruce and Leo exited Willcox and got into the desert, Leo sat up and verbally commanded the AI.

"AI. How far is the Steward observatory from the nearest residential area in Tucson?"

AI responded right away. "The residential area off Silverbell Rd, Tucson, is 3.2 km from the Steward Observatory. The time to travel on foot from the Silverbell area to the Steward Observatory is 40 minutes for an above-average physically fit person. It continued, "There is another residential area to the South of the Steward Observatory..."

Leo waved his hand to stop the AI from giving results because, chances were, it would continue for some time.

"So?" Bruce looked at Leo, "Road or no road?"

Leonardo shifted in his seat and checked his watch uncomfortably. They both knew he wasn't in the best physical shape and that walking the road to the observatory would be challenging, let alone hiking up the mountain trail from the South.

"No road," said Leonardo. "And here's why."

The two men spoke for the rest of the way to the mountain base where the Steward Observatory sat. It was mid-morning when they prepared for the eight-kilometer walk to the observatory. They didn't know that while they were on their way to make the Steward Observatory safe, another person from their time was traveling in the opposite direction to Willcox.

Chapter 27: The Life of a Killer

Shadow sat motionless at the back of the Greyhound bus as it wound through the desert landscape. To the exterior, Shadow looked calm and relaxed; however, inside, Shadow was going over plans, dates, and locations, ensuring they were all committed to memory. Being prepared would guarantee success in the mission, as it had many times before.

As driven and self-motivated as Shadow was, a place not far under the surface was severely damaged psychologically. When only 13 years of age, the lure of money attracted the young Shadow like iron filings to a magnet. Growing up in a brutal atmosphere where gangs ruled the streets, the person who was willing to do what all others wouldn't was the person who rose in those gangs the fastest. Had Shadow not been recruited by a covert division of the Army at just fifteen years old, death would indeed have been Shadow's fate.

Known only as "Alpha 6", the black ops unit that trained the young Shadow used the youth's intuitive killing nature for many assassinations worldwide. The 6 in Alpha

6 referred to the old military jargon for watching someone's back. Alpha 6 watched the back of the world's superpowers, and its covert operatives carried out killings when the Organization deemed a person or people a threat to world security.

At first, Shadow believed that killing for the military would bring honor and make a positive difference in the world.

Soon, killing was second nature, and being an Alpha 6 covert operative was the only way Shadow knew how to live: solitary, off the grid, with no fingerprint and no relationships that were unnecessary to the mission. Shadow did what was necessary to infiltrate initially but ultimately became addicted to the hunt.

Head shaking to focus on the task, Shadow mentally reviewed the mission waypoints, the equipment stored in the duffle bag for the mission, and the map with street names that had been committed to memory so long ago. Shadow's watch held all piecemeal information; however, depending on 22nd-century technology was not a crutch Shadow wanted.

The Greyhound bus slowly squealed to a stop as Shadow got up and grabbed a black duffle bag from the empty seat to the side. Walking heel to toe with deliberate steps, Shadow automatically balanced in a military-style stance as the momentum of the bus changed, and it slowed to a stop.

When the bus roared away in a cloud of black diesel fuel exhaust, Shadow slung the duffle bag and began walking. *The best defense is an offense,* Shadow thought as the buildings came and went, and the people went about their daily business in Willcox, oblivious to the killer's presence, let alone the epic battle that now had its epicenter in their town.

Chapter 28: Little Franny

From a young age, Fran O'Shea was naturally inquisitive, so she had been drawn to science and dedicated her life to it, particularly particle physics. Her friends, who came and went as she breezed first through elementary and secondary school and then rapidly progressed through multiple university degrees, had nicknamed her "the bulldog" because once Fran had dug her teeth into something, she never let go until she found what she was looking for.

As is often the case in experimental physics, the answer is, 'This is not the solution you were looking for.' That answer never stopped Fran for a moment. By the time she finished one scientific experiment and documented the findings in a scientific journal, the following experiment and the subsequent theory were already in their infancy in her mind.

Born in Scotland in 2099, Francis O'Shea demonstrated a high aptitude for mathematics at a young age. The academic community in the 32 Nations group, which included Scotland, had established early testing for children who demonstrated extraordinary intellect in many areas, including mathematics. Her parents were both elementary school teachers in the small rural community of Wigtown, Scotland, when they completed a request for their 3-year-old daughter Fran to be tested related to her mathematical problem-solving skills.

It took the tester 3 hours to test young Franny via a holographic conference, but he knew her massive potential within the first 10 minutes of meeting the young Scottish girl. Her mathematical score, which was recorded that day under the World Academic Testing Standard (WATS), was promising, and her parents were spoken to by a professor of mathematics and physics from the University of Geneva within a few days of her testing. The results that Franny had achieved while she was 'playing a game' with the tester showed that her propensity for problem-solving and primarily abstract mathematical thinking was in the top 1% of gifted children tested in her age group.

Within two weeks, Angus and Mary O'Shea were offered teaching jobs at an elementary school in Geneva with the understanding that little Franny would be attending the child enrichment schooling program at the University of Geneva five days a week. The university also sweetened the deal for Franny's parents by offering them free room and board at one of the family suites on the university campus where Franny would be attending school. Everything was within walking distance or a short bus ride for the O'Shea family, and the university hosted many cultural, musical, and entertainment events meant to stimulate the young, brilliant minds of promising little people.

Having few living relatives in Scotland at the time, the O'Sheas sold their Wigtown house, car, and all other items they wouldn't need and left for Switzerland. The parent community at Geneva University was interwoven by design. Angus and Mary found some like-minded friends, including a couple who had moved to the university from Scotland with their son five years earlier.

The environment was peaceful, stimulating, and engaging, and little Fran O'Shea excelled at her university studies.

By the age of nine, Franny had completed all the required tasks and requirements that were considered high school level, and by the age of eleven, she had completed an undergraduate degree in Physics. By then, Franny had grown tired of her nickname 'Franny', so she insisted that people call her Fran, except for her parents, who always called her Franny when they spent quality time together.

Fran had always been interested in particle physics, and, at the encouragement of the university, she began a Master's degree in particle physics at the age of twelve, where she met doctor Leonardo Hoffman, who had come to be a guest lecturer on Quantum Mechanics and time travel theory. After the lecture, doctor Hoffman sought out young Fran O'Shea and spoke to her and her parents about an opportunity for Fran to study at CERN, which was only a fifteen-minute shuttle ride from where they lived at the university.

He proposed that she complete her Master's and doctorate degrees while supervised at CERN. The young Professor himself was not even 30 at the time. Although Fran had begged her parents to allow her to study at CERN, they ultimately declined, citing that she was too young to work with people, many of whom were double her age. They had agreed that they would take her to CERN to visit and allow her to take her Master's degree in Geneva, which was now home, and after that, she would be old enough to do her doctorate work at CERN under Doctor Leonardo Hoffman.

Doctor Hoffman politely accepted Fran's parents' decision; however, he ensured that one of his colleagues at Geneva University would' steer' her in the right direction so that transitioning her studies to CERN would be seamless when the time came.

While at Geneva University, Doctor Hoffman would occasionally visit Fran and her family and take them out for dinner. This kept the communication between the doctor and Fran's parents regular and solidified his presence as a family friend. The O'Sheas always looked forward to the remarkable mathematical conversations that Leonardo and Franny would have across the dinner table.

It was a productive, peaceful, and quiescent life for Fran and her family.

Chapter 29: The Forensic Report (11.03.2135 20:00 hrs.)

Fran stirred the sugar into her tea, watching it dissolve as she slowly moved the spoon back and forth. The complex partial differential equations that determined the movement of the sugar and tea as they swirled around to combine and create a relatively homogenous drink came to her mind as she stared at the tea and watched the eddies in the liquid appear and disappear. She and Mitch had worked in the same office all day and had hardly spoken to each other the whole time.

"So," Fran was still staring at her tea. "I still don't understand why we can't see the forensic report from that attack on the lab. Honestly, Mitch, doesn't that sound suspicious to you?"

"Well, uh…" Mitch was staring at his shoes. *Again.*

"Our team is decimated, and the mission compromised; Mitch, you better stop looking at your shows and start talking, or I'm going to throw you right out that door and go it on my own!" Her native Scottish accent came out most noticeably when she was mad.

She fumed, "The board of directors seems to be floundering; what the hell are we doing here if we can't know what happened to Leo or find out who killed Gus?"

Mitch cleared his throat, "well, they do have that guy in custody."

Fran's eyes got larger than usual as she fired back, "Mitch, give your head a shake! I've looked into that guy they have in custody; he doesn't seem smart enough to steal a car, let alone break into one of the most highly secure facilities in the world. Oh, and make Leo disappear off the face of the Earth at the same time. All we know is that there was possibly two people's blood in Gus' office and the hallway to Leo's office, and there was a bullet hole in Leo's wall."

Fran was winding herself up; the next part got even louder and even more Scottish, "Oh, and let's not forget about that super creepy phone call I got from the director. What the hell was in Leo's desk drawer that is missing?"

Fran stopped, took a deep breath, and closed her eyes. I wish I were *back at school in Geneva, where everything was normal.*

Mitch lifted his head from his defeated position and spoke, trying to calm his partner down.

"Fran, all we can do is follow what they tell us. Leo was in charge, and now I guess the board of directors is in Leo's place, and they'll tell us what we need to know."

"You're such a coward, Mitch," Fran shot back, "are you seriously going to just bury your head in the sand over this? One of our coworkers is dead, and another is missing. I'm pretty sure we've been given bogus information, and they have this stooge in custody; there's a bigger picture here we aren't being allowed to see."

There was a long pause where both Fran and Mitch were silent. Fran broke the hiatus again.

"You know, I bet they're recording us right now," she poked at when Mitch didn't respond. "They're probably recording us at home too; recording your family, Mitch."

"Okay, okay, Fran. What do you want me to do? Mitch had his hands up, surrendering to her barrage. "I'm sure we're under constant surveillance, and it freaks the hell out of me. I say we wait for instruction from the board of directors and go from there. That's what's safe."

"Ohhhhhhh!" Fran yelled, exasperated.

She stood up, took her cell phone out of her pocket, and threw it in the garbage can. It gave a loud thud as the phone bounced off the inside of the garbage can, followed by a loud bang as it hit the bottom of the can.

Looking around her and up into the industrial wiring, panels, and electrical components above the lab, she yelled, "Okay! We know you're watching us! We know something is up because we don't believe for a second that - that moron you have in custody did all of this! Answer me!" Fran sat down, flattened from her outburst at the space around them.

Fran and Mitch sat silently for well over a minute before the silence was pierced.

The phone in the lab rang.

Mitch and Fran froze and looked at each other as if to ask who would answer the phone. Knowing Mitch was too chicken to answer, Fran tentatively walked over to the phone and slowly picked it up. It was an internally wired phone, so only calls to and from the HEPA lab, nearby offices, and security could be made.

She paused for a second.

"Hello?" The faint, high-pitched digital squeal of an encrypted audio message rang in Fran's ears. She spoke again, "hell-." The following sounds cut Fran's voice off, and as she listened intently, her eyes widened with a look of sheer bewilderment on her face. She listened for about twenty-five seconds as tears gathered in her eyes. Then, without a word, she hung up the phone and sat on the floor, burying her face in her hands as she began to cry.

"What the hell?" Mitch exclaimed. He got up to go over to Fran, but she put up her hand for him to stop.

Fran got off the floor, wiped the tears from her eyes, and, with a deadpan face, said to Mitch, "My godfather passed away today."

Mitch froze, his mouth open, his eyes darting around, not knowing what to say or do.

Fran spoke again but slower, "I have to leave. Mitch, my godfather, passed away today."

Mitch sat down on the nearest chair and nodded. He

couldn't get any words out of his mouth until Fran was almost out of the lab doorway.

"My condolences!" he croaked; his mouth was so parched. He swallowed. His own emotions were welling up from all the events of the last couple of days. Mitch put his hands on his face and started to cry; the emotional toll of the attack on the lab and what had just happened wholly overwhelmed him. As he sat with his hands on his face, his thoughts were racing at the possibilities of what had just happened.

My godfather passed away today, was one of the phrases that Doctor Leonardo Hoffman had insisted Mitch ingrain in his memory because it meant that they had received an encoded message from Leonardo that the HERB mission was compromised.

Mitch now had a series of tasks that he had to complete over the next few days due to those code words being uttered. He went through each one in his mind: the locations, the timings, the tasks, all while he had his hands over his face. He couldn't allow any suspicion to arise to those people who were watching their every move. *Things just got a whole lot more serious.*

As Doctor Fran O'Shea walked to the bus terminal in the rain that night, Leonardo Hoffman's digitized voice still rang in her ears. *Your godfather passed away today. Was Leo alive? Where the hell was he?* All she knew was that if her fictitious godfather had passed away, she would do a unique series of tasks and then wait for further instructions.

She patted her pocket where her cell phone would usually be and remembered, with an "Oh no!", that she had thrown it into the garbage can during her adult temper tantrum in the HEPA lab.

She had done that as an act of rebellion, knowing she was being monitored, likely via her cellphone when she wasn't in the HEPA lab. She would have to return in the morning, retrieve it, apologize to Mitch, and tell him that he was right and that they should wait for instruction from the board of directors regarding the mission. She didn't believe that for a second; however, the new development that there was a message from Leonardo himself that the HERB mission had been compromised changed the whole game plan.

If Leo was alive or could get a message to them,

they could figure out how to proceed. Gus had been Leo's unofficial right-hand man, and as Fran went over the details of Gus' death and Leo's disappearance in her mind, it made sense that if both men were removed from the mission, Fran and Mitch would be left floundering. Not floundering from a lack of intellect or anything else, just a lack of information; the HERB mission was shrouded in layer upon layer of secrecy.

Fran and Mitch were told what they needed to know. *So was Gus, for that matter*, she thought. Fran got off the shuttle that stopped within a stone's throw of her home and didn't even bother moving the wet hair stuck to her face until she got to her front door. The door required facial recognition to unlock, so she swiped the wet strands of hair away from her face with her hand and looked at it. And after she heard the deadbolt disengage, she walked in.

After removing her dripping jacket and shoes, she sat on her couch. She began to cry again, looking at the small cabinet to the side of the room with pictures of her mother and father. She thought, *I wish you were still alive*, then instantly revoked the thought. *No, I would drag you into this mess if you were still alive.*

Fran leaned back on her couch and took some deep breaths. She started to think about Leonardo; it was the time that he had come to meet her and brought her out for a cappuccino in the busy Swiss marketplace near her home.

They sat outside at one of the small, round tables at the café. It was the day before she was to start her new post-doctorate project with him at CERN. It was a beautiful, cool, sunny Swiss day. It was July 5, 2131. He leaned in to articulate the information he was telling her quietly.

"Do you understand what I'm telling you, Fran?" Leo looked at Fran's eyes intently. "This will probably never happen, but I need you to commit the phrase and sequence of tasks to memory. Never record it in any way; keep it up here." Leo pointed to the side of Fran's head as he leaned in to speak. He put his hand on her face and continued, "Should something go wrong with this mission that you are now a part of, you will receive that message in my voice. Do you understand?"

"Yes, doctor Hoffman, but I don't understand why..." Leonardo cut her off.

"You just have to trust me, and I must be upfront

with you before you start working with me on this mission; there is a lot at stake. More than you can imagine right now."

"Fran smiled. Like what?" she looked at Leonardo, who did not smile back; he was dead serious.

I've been working there for years; what could he be up to? Fran thought as she studied Leo's face.

Leonardo continued, "I can't tell you until we are in the lab tomorrow, but you can tell nobody nor speak about this conversation, even with me. Ever. The others you'll be working with within the lab will know their roles in a fail-safe protocol, but you must understand that your role is the most important, and you cannot deviate in any way from these instructions."

The two Physicists quietly talked over their cappuccinos while the busy marketplace bustled around them. As doctor Leonardo Hoffman got up to leave, he grabbed his full-length brown jacket and put it on, then leaned down to kiss his colleague, as was European tradition. As he leaned in, and before he planted the friendly kiss, he whispered, "Commit it to memory. Today." Doctor

Hoffman walked down the busy street and got onto the shuttle back to CERN.

Fran returned to the present, got off her couch, walked to her bathroom, and started the shower. She was nothing if not a good student, a good scientist, and a good person, and she had committed Leonardo's instructions to memory when he had told her almost four years earlier, and she had not forgotten one word. Her memory was like that; she could be given a piece of information or a formula once and recall it exactly years later. As she showered, she mentally reviewed her course of action over the next several days.

By the time Fran had finished her shower that night, all the electronic data on the passing of Fran's godfather were automatically uploaded to a Scottish local news site, complete with photos in his memory, some comments from his friends and family, and a death certificate—all fake.

Unknown to anyone on Earth in the year 2135, there had been a pre-programmed subtle frequency transmission made when doctor Leonardo Hoffman logged onto his computer three days earlier; there was a gun pointed at his temple, and he had already been shot once in the shoulder.

The highly encrypted signal that was transmitted through the low voltage internal phone line and then was piggybacked through the CERN video camera power system undetected was a pre-determined message. It had been one of the many fail-safes that Leonardo had put in place should things go wrong with the mission.

He was a man who spent his entire life analyzing, planning, and thinking; multiple levels of fail-safes for a critical mission, such as the HERB mission, were warranted and were set up in several different electronic and non-electronic media. Leonardo recorded holographic video messages for as many different scenarios as possible; being the only connection between the board of directors and his team, he knew that should something go wrong and he could not contact them, they would be lost in the dark.

He knew that if the HERB mission had somehow been compromised, he would have to be able to pass along information to his team in secret.

They were the only people in the year 2135 that he could trust completely with the destiny of the mission.

The electronic frequency transmission that had been

sent was triggered by the AI on Doctor Hoffman's computer, which had detected that he was under duress. It knew normal pupil size variations, body temperature, breath rate, heart rate, sweating, and blood pressure, and it detected the visible injury on the side of his face. AI also knew the doctor would never input passwords on his computer with another person in his office.

A simple scan, invisible to the human eye, was done by AI when he sat at his computer. The series of words auto-populated on the computer screen when AI saw that only Leo was watching the screen; they were a pre-determined series of words and characters that would be meaningless to any other person; however, when Leonardo looked at the red code words displayed on his computer screen, and as he was mimicking typing those words, he was confirming the type of emergency that AI had detected.

Fractal   Eigen   Apple   Tetanus   Einstein   Mandelbrot#

In those moments before time traveling to the past, he had set the wheels in motion that could not be stopped, diverted, or altered; he had confirmed the worst possible scenario. It could not be undone. Deep within the encrypted bowels of Leo's computer, AI had executed the program, *Fractal Eigen Apple Tetanus Einstein Mandelbrot#*, then immediately after that, had wiped its entire memory completely clean.

Years earlier, as Doctor Hoffman wrote the same commands that were just executed, he thought that an opportune quality of AI was that it didn't have any feelings on the matter, nor was it capable of deviating from what it was programmed to do. He hoped at the time that none of his fail-safes would ever have to be used, and he never told another person about the full scope of the safety nets. He programmed the fail-safes that the board of directors had recommended as part of the HERB mission; however, deep within the source code of the HEPA server, he embedded his fail-safes that superseded theirs to guarantee the fate of the HERB mission.

Chapter 30: You've Got Some Explaining to Do (29.11.1981 08:25 hrs.)

"So, Leo, why are you insisting on going this way again?

It makes no sense, and in your condition, no offense, but it makes way less sense." Bruce was starting to get impatient.

Leonardo Hoffman was sitting comfortably in the passenger seat of the '76 Impala as they gas-guzzled their way toward the South of the turnoff to the Steward Observatory. He had an air of smugness and a smirk on his face that was uncharacteristic of the scientist.

Bruce playfully raised his voice, "Leo, seriously! You need to fill in the mission specialist with your smug little thoughts because I don't know what you're getting at going this way. Should I ask AI?"

"HA!" laughed Leonardo, "For once in a long time, this AI doesn't know either. It's such a relief to know that in this old, beat-up, ancient gasoline car, that stinks, by the way, nobody is listening to me but you and this standalone AI. There's no internet or cameras, so we are free to say whatever we want in this car right now. Before I came here, I hadn't experienced that in my adult life, and it is absolutely refreshing." Leo put his head back on the oversized leather headrest and smirked for a moment longer before explaining.

"I won't keep you in suspense any longer, Champ." Leo smiled like a fool, looking at Bruce. "Stop over here on the side of the road for a moment."

Bruce slowed the vehicle and stopped on the side of the road; the loud tick, tick of the signal light, and the gurgling engine were all they heard for a few moments while Bruce stopped the car and put it in park.

"Okay, what?" Bruce was now irritated with the lack of information, but as they stopped, Leo smiled as he powered down the AI. Bruce knew this would be a doozy.

Leo started his long-awaited explanation. "Bruce,

do you remember all the fail-safes planned for this mission?"

"Uh, yeah, Leo, I spent a painfully long time memorizing them before my final interview. I can list them by code number or alphabetically if you like."

"Well," Leonardo smirked, "I made some of my own. In fact, I created one hundred twenty-eight of my own fail-safe procedures and embedded them deep within the HEPA lab server. So deep that the director himself didn't know about them." Leo smirked. "And that big, bossy, know-it-all all prick can't hear any of this!" Leo laughed again after his sarcastic remark, joyous with the lack of technology around him.

Leo continued, "and I kind of programmed my one hundred twenty-eight fail-safes to override the board's eighty-five fail-safes for the mission. I also allowed them to believe that their fail-safes were the only ones on the server, and they will appear as though they were executed, given what happened in the lab the day I transported here." Leonardo opened his arms wide at the empty Arizona desert around them as if to take a bow.

Bruce couldn't help but smile. "Okay, Leo, you tricked them all. Now, can you tell me what we are doing on the South side of the mountain, stopped on the side of the road, and why I'm listening to you gloat about how smart you are."

Leonardo straightened up for the next part. "One of the fail-safes I created, the one that I executed right before I transported to this desert, has set in motion a series of events that will very shortly become a great advantage for us."

Bruce was getting frustrated at the cryptic delivery, "Okay, Leo, out with it. How could one of your fail-safes become our advantage when we are both in 1981, both time travel devices are also in 1981, and one of the devices has been stolen?"

Leonardo smiled, "Do you remember the first time I showed you the mouse that I transported in time by about ten seconds?

"Yes, of course, Leo, I remember. You don't forget the day when time travel is revealed to you, and a mouse disappears and reappears right in front of your eyes."

"Well, that same experimental testing device is not capable of transporting any human, nor anything larger than a shoe box, but it is still much more functional than anybody knows, and by yesterday evening, a package no larger than a shoe box was transported from the year 2135 to 1981." Leo looked out the window, smiling.

Bruce was shocked. "You kept this from me the whole time?"

"Yup," was the retort from the scientist, who was still smirking out the window.

"What's in it, smarty pants? Why didn't our HERB device alert us to a time travel event? Huh?" Bruce was now smiling at the fortuitous turn of events.

"If my program worked undetected, as I'm sure it did, we can park up ahead where the hiking trail starts, and I'll show you where it is. As for the HERB device being able to detect the new time travel event, I took the liberty of making it a completely standalone device. Besides, it's not quantumly entangled with the two HERB devices in 1981."

"You took the liberty, eh?" Bruce reached over and

hit the scientist on his forearm, then squeezed it as he grinned ear to ear.

Bruce and Leonardo walked silently up the hiking trail; Bruce's backpack carried the AI and all the other equipment they needed. Having seen no other person since they had parked the car, they veered off the hiking trail towards a large rock outcrop that was still due South of the observatory and well below the summit of the hiking trail.

They entered a small cave that was obscured from view from the trail and located the black shoebox-sized container that was sitting at the bottom of a pool of water that was about half a meter deep. As Leo gingerly opened the 22nd-century container and after drying off his hands and arms, Bruce's eyes lit up like it was Christmas.

"We certainly have the upper hand now," Bruce crowed.

Shortly after, Bruce and Leo separated as Bruce continued the rest of the way up the mountain trail to the summit where the Steward Observatory stood. Leonardo stayed in the cave and watched Bruce's every move.

The multi-use robot programmed to find one

particular shoebox-sized container in storage and bring it to the testing area of the HEPA lab had activated the program to transport the container, returned to its docking station, and powered down. As per the programming overriding its normal controls, it was not transmitting any information about its whereabouts, activity, or the programming commands it was given. The robot then silently executed the program that would wipe out its memory of the most recent actions it performed. It would then appear to the maintenance people working at the HEPA lab that the robot had had a fatal breakdown in programming in its circuits.

As per HEPA lab policy, the robot would then be wiped clean of stored data and disposed of. The clever series of events that Doctor Hoffman had conceived of years earlier was completed without the knowledge of any other person in 2135.

Even the large amount of energy required for the object to be transported back in time had been steadily trickling into large energy-storing capacitors for days to eliminate any noticeable surge in power consumption in the HEPA lab.

The transmission message from Doctor Hoffman's

computer to the robot that carried out the deed was kept on the internal server in the maintenance band of information that was constantly flowing around the laboratory. It had been ping-ponged around to several different machines in many fragmented pieces before arriving at various times at the selected robot, where the message was reunited.

One piece of the puzzle was added at the end to bring all of the electronic bits of information together to execute the failsafe protocol, courtesy of one of Leonardo's research team, who was unknowingly giving the final command. Finally, the burst of gamma radiation from the instant the Hoffman-Einstein Rosen bridge was opened in 1981 did not register on any instrument in the lab since the experiment room was shielded and away from the rest of the lab.

In short, Doctor Hoffman counted on that unless someone knew the exact details of the fail-safe protocol he had put into place or knew enough to follow the maintenance robot moving surreptitiously about the lab and executing the commands, they would never know what happened.

B.E. Smith

Chapter 31: Mission Damage Control (11.03.2135 20:30 hrs.)

The director walked into the HERB mission board room deep underground and immediately took control of the room. He had a palpable gravitas about him, and all the board members quieted. He could hear more than a couple of gasps from the five men and five women talking nervously in small groups around the room.

The director could see holographic flow charts and lists over the center meeting table; however, none of the eyes were on them. All eyes were on the director. He rarely attended in-person meetings; when he did, he commanded the room like no other board member could.

"People, please sit down," the director announced, waving his arms out and dipping his scarred, bald head to show all the empty chairs around the table. All the board members went to their usual seats as the holographic projector shut off and the surrounding lights brightened; there was complete silence as the director used the retinal scanner to confirm his physical presence at the meeting and spoke in a soft, deep voice, "holographic file - damage control 1 to board room projector."

The side lights dimmed, and the holographic projector started by displaying the 3-dimensional crest of the HERB mission. They had chosen a simple graphic at the onset of the mission: a rainbow bridge over the top of the Earth with a small black sphere above the bridge. Everyone else on Earth assumed that the small black globe on the HERB crest represented the asteroid, but it depicted the HERB time travel device itself.

All the board members' eyes were on the director; he detested being around people; however, he did love the feeling of being in control, and he was in control of the boardroom.

The director looked around the table to the other ten men and women before he spoke.

"Approximately eighty hours ago at the HEPA lab, an unidentified assailant murdered doctor Gus Maxwell and abducted doctor Leonardo Hoffman.

We do not know Doctor Hoffman's condition, status, or location. As a result, we do not know if, or more likely, how many critical mission secrets were compromised. People, the HERB mission has been compromised. As you all know, as a part of this mission protocol that we have all collaborated on for years, all of you have submitted, including me, to an AI forensic interview to determine if there has been an internal leak related to this critical incident in the HEPA lab. AI has analyzed all the interviews and concluded that no internal leak was sourced to this board of directors. For that part, I am thankful but not surprised."

The director paused and looked around the table at all the relieved faces. He sighed with frustration before he continued, "The forensic report," as the director waved his hand towards the middle of the table, and the holographic projector silently followed, "has been reviewed personally by myself, and I will provide each of you a copy of this report to view in this room only. No explanation is needed as to why that must be. We are constantly assessing this fluid situation with every resource we have. Still, the fact remains that an unauthorized individual got into the secure HEPA lab undetected and, once inside the lab, fired three rounds from a magnetic railgun. This individual then somehow escaped," he clenched his jaw as he continued, "and in the process, abducted Doctor Hoffman."

Maryam, one of the female board members, spoke up, her face red with indignation, "But that is absurd! How is this even possible? The HEPA lab is an ultra-high security state-of-the-art facility!"

The director put up his hand in agreement, nodding his head.

"I know it is baffling to all of us, but as you know, we are looking at every detail, no matter how small, to attempt some kind of damage control. Our most sophisticated AI programs have been running 24/7 since this incident."

As the board of directors watched, the holographic projector cycled through all the security means at CERN and HEPA and what they showed during the breach and the subsequent escape. None registered anything out of the ordinary. AI began showing brief transcripts of video statements from security staff, maintenance staff, and delivery personnel. All had appeared to be truthful, and AI had determined that none of those interviewed knew anything about what had happened.

The director said, "We have set the parameters for AI to search for anything out of the norm in the last two weeks before the attack and up to our current time. And if nothing helps us, AI will begin scanning backward in ten-day blocks at a time."

Pieter, another of the board members, spoke up, looking around at the other board members as he spoke, "What about the man in custody? We were led to believe that it was just a matter of time before Doctor Hoffman would be returned, and we could move on with the mission." Pieter lowered his head, "and may Gus Maxwell's death be mourned once we are in a position to do so."

The director nodded, "Yes, it is a tragic death; Gus Maxwell was a valued and integral part of this team. The man in custody has committed crimes, but unfortunately, he was not involved in this attack, Gus' murder, or the kidnapping of Leonardo. Not directly. We must give the illusion of closure to the media because they cannot know that there has been a severe blow to this mission; public opinion is high that the HERB mission will succeed, the asteroid will be diverted, and we must allow them to keep believing that. The world thinks that we are diverting the asteroid now, in this time, with new propulsion technology from the HEPA lab, for crying out loud! We need them to keep thinking that."

The director stopped and looked at each of the faces around the table.

He raised his hand and pointed his finger at the holographic video in the middle of the conference table. As the director began to speak about the asteroid, the video changed to a graphic of the asteroid's path and collision course with Earth.

"We are at the most important juncture in the history of our species, and we will be victorious over this saboteur. This mission will save the Earth. You have my word on that."

Another board member, Katherine, interjected, "Years back, as we prepared for the original HERB mission, we spoke of the twin time travel device created as a failsafe if the primary device malfunctioned.

Our knowledge at the time was that the two were quantumly entangled. Is that device secure, and can we use it to ensure the current mission's success? Can we send the other candidate back using the second device to assist Bruce?"

Nodding, the director conceded. "Yes, the second HERB device."

The director pulled at his goatee as he stared at the holographic images of the asteroid over the table, "Unfortunately, it was temperamental and unreliable in testing, so its HERB capabilities were decommissioned right after this last mission was commenced. Currently, all that is possible for that device is to send an emergency communication to the primary device and recall Mr. Hayden. We cannot return Bruce at this juncture; it would mean suicide for this mission, and we don't have enough time to restart it again. As we are all painfully aware, from the known physics of this kind of time travel, traveling back and forth to the same time and place weakens the folded structure of the Hoffman-Einstein Rosen Bridge, which is why we have sacrificed two years in our time each time Bruce has attempted this mission."

He continued, "Unfortunately, we are out of time, so as far as the ongoing mission goes, besides doing what we can regarding damage control on our end, we have to trust that Bruce will be successful. The part of the mission that is here now, we will deal with it most thoroughly and sensitively; I'm sure you can appreciate how sensitive you must all be to this information. People, this is our last stand."

He scanned the faces around him again, "Let me reassure you, the mission codes that could be used to retract Bruce Hayden from his mission have not been compromised."

Maryam bolted out of her chair, "Director, you didn't answer Katherine's question. Is the second HERB device secure?"

The director looked at Maryam and clenched his jaw; his muscles flexed before he spoke again, "The second device is missing along with Doctor Hoffman."

The board of directors table erupted in gasps and shouts and then side conversations; some leaned back in their chairs, some stood up, raising their arms, and some hunched over, putting their heads on their arms on the table.

The director raised his voice, "People!" The noise settled slowly, but everybody looked stressed.

"People," the director continued, "Sit down, please, and listen! I believe we still have a window of opportunity here to solve this, and I can tell you with great certainty that if Doctor Hoffman is alive, he could not know the mission codes if he is outside this facility, as we know he most certainly is.

There is no way he is still in this facility, and in that way, the mission is still safe. Let me reiterate to you all that it has been confirmed that none of the mission codes were leaked during this attack, and the codes change every twelve hours; therefore, we can only assume that whoever has the twin HERB device is holding onto a multi-billion dollar, glorified eight-ball. All the forensic reports from the technological standpoint confirm that there was no mission code breach. All we can do is hope that Doctor Hoffman and Doctor Maxwell didn't both sacrifice their lives to maintain the integrity of those mission codes and that Bruce pulls through like we expect him to."

The director stood up and left without another word as the holographic projector dimmed, the surrounding lights brightened the surrounding board room, and the board members continued to voice their indignance at the stolen HERB device.

Before long, the director was back in his home workspace. He sat in his chair and sighed.

His hand movement called up the file related to Doctor Leonardo Hoffman's office computer, and the director stared at the results of the forensic analysis and AI interpretations. He scratched his head and stared at the data. *What are you playing at, doctor Hoffman?* he thought, as he lost himself in the reams of holographic data around him.

Chapter 32: The Steward Observatory (29.11.1981 9:55 hrs.)

Leonardo sat comfortably in the cave on top of the shoebox-sized futuristic container and watched the video stream from Bruce's nano cam. He used it along with the integrated AI to scan continuously in its field of vision. Bruce swiftly walked up the mountain trail; his nimble feet navigated the tree roots and rocks with expert precision.

Bruce was engrossed in the mission that afternoon; he knew a lot was at stake in making the Steward Observatory safe for him and Doctor Nault, so they could conduct their research in the short term until Leo could arrange a change of observatories.

"I'm coming up to a secondary plateau here, Leo," Bruce whispered as he stopped just short of the little clearing to allow the AI to scan and for him to visually inspect for any sign that Shadow had been there.

Leonardo chimed in Bruce's ear, "Looks okay from

what AI is seeing, Bruce."

"Okay, I'm going to continue," Bruce whispered, "but I think I'm going to keep close to those shrubs as cover and not get into the open space ahead."

"You're the tactical wizard," Leo responded in Bruce's ear.

Bruce scanned the entire landscape of the open area; he saw some older footprints that had scraped mud over one of the small rock outcrops, but he didn't see anything suggesting that Shadow had been there. *He knows the same as I do regarding the tactical movement in this terrain, if not more,* Bruce thought as he walked along the shrub line to circumnavigate the open area.

There was a noise off to the side of the trail and off to his left in the tree and shrub-covered area. Bruce held his breath and swiveled his head to the left without moving his body while he listened and allowed the nanocam to capture the source of the noise.

"It's just a skunk about ten meters to your left," Leo whispered into Bruce's ear. "Don't get sprayed because you'll have to walk home." Bruce could hear Leonardo chuckle.

Bruce continued to walk along the bush line and had about fifteen meters left before he met up with the partly covered trail again.

"Hey, Mr. funny man, how about we focus on getting this job done?" Bruce was a little irritated at Leo's seeming lack of focus; however, he was still delighted at the distinct advantage they had just gotten with the arrival of the package from the future, courtesy of Leonardo Hoffman.

The nanotechnology was impressive: earpieces, a nano drone, cameras, injectors, and weapons; seemingly, the doctor Hoffman that Bruce knew was not opposed to any of the 'nano-no-no's' that Bruce had asked for before this mission. Granted, he had taken advantage of the one weapon they gave him in his last assignment: the injector. It was called an NDI, which stood for Neurosynaptic Dampening Injector. It was a nice, friendly name for a disabling nano-injector that took the unlucky recipient entirely out of the equation.

He had used an NDI on his stalker, Theresa, in his

second mission attempt when she somehow found out where Bruce and Kevin were staying and had come to tell Bruce about the baby they had together, which didn't exist. She had threatened to kill Bruce on more than one occasion, so Bruce had surreptitiously deployed the injector, and Bruce's Theresa troubles ended.

Theresa became a zoned-out madwoman, incapable of thinking any rational thought process and incapable of stalking Bruce or Kevin anymore. After several months, the unexpected side effect of the cocktail on the woman's body was that she became homicidal and killed four innocent people, hitting them with a van that she had stolen.

Bruce thought it was a proper use of the nanotech at the time; however, the board of directors deemed that Bruce's possession of these instruments was more of a concern than they were worth. Hence, Bruce was not allowed any more NDIs on his missions. That is until his friend Leo conveniently brought them in his checked luggage from the future. *Desperate times call for desperate measures,* he thought.

Bruce came up the final slope towards the peak of Mount Lemmon, which towered over the rest of the area around Tucson, and saw the observatory there. It was two kilometers higher in elevation than the rest of the surrounding area; it was a beautiful sight around him, but he had not examined every little detail around him to determine if Shadow had beaten them to this place.

A tall, white-domed building that crested the mountain was not exceptionally spacious after the precious telescope's moving parts and mirror occupied their necessary space in the building. As far as workspace inside the observatory, what was left was taken up by a few small offices and a kitchenette for the people who had procured the telescope's time in blocks of a few hours up to days at a time. Besides knowing when Doctor Nault's telescope time was and crossing paths with other scientists before and after the nighttime block on Tuesdays and Saturdays, Bruce didn't take the time to know who had what blocks of time and when.

Once he and Kevin had made their discovery, the telescope would be theirs whenever they wanted; however, now that Shadow had put a monkey wrench in the mission, they would most likely be making their initial discovery of NH-1983 from another telescope, quite possibly in another state entirely.

After Bruce made his initial observations of the open area leading up to the telescope building and confirmed with Leonardo via their new communications link that AI detected nothing out of the ordinary, he took a deep breath, bounced his packsack up onto his shoulders, and began walking up the roadway to the observatory.

One vehicle was parked outside the observatory building, which meant that one or two scientists were likely getting ready for their nighttime exploration via the telescope. Bruce would waltz into the observatory, posing as a student hitchhiking through the area and exploring the mountain. He needed to place the one weaponized nanocam inside the building, and now, thanks to Doctor Hoffman, he would put an extra camera outside on the grounds.

AI had already given the optimal locations for both, and Bruce, who knew the telescope building inside and out, would place the cameras and get out of there. While walking past the large, well-trimmed bushes surrounding the telescope building in a pleasant architectural design, he heard Leonardo yell in his ear, "Stop!"

Bruce froze and instinctually adapted to a combat stance, not knowing what was happening. After a few pregnant seconds, he whispered, "Leo, what is it? You better not be jok-." Leonardo cut Bruce off, and Bruce braced himself for what came next.

Leonardo's voice was panicked, and he spoke deliberately slowly, "Bruce, off to your right, about two o'clock, AI picked up a high-frequency transmitter. It's in the 8GHz frequency band, possibly a nanocam."

Bruce had backed himself up between two large, tall bushes, with two or three brushes between himself and the object Leo was talking about. He knew that if the thing transmitting were a nanocam, it would be able to see him day or night, bushes or no bushes, in multiple different wavelengths.

Leonardo Hoffman spoke again as Bruce could hear him furiously entering commands on AI, "Bruce, I am pulsing your nanocam frequency in several directions around you so that if there are nano-weapons up there, it might scramble their lock on you."

"Great," Bruce whispered. "I'm looking right in that

direction; can you see anything else to guess what exactly we are dealing with?" Bruce carefully removed his packsack and placed the tiny canister containing the two nano cams in his pocket; the rest of the contents of the bag were expendable, should Bruce have to run and throw his packsack as a decoy and avoid whatever weapon that device had.

Leonardo responded, "Just stay perfectly still, champ. I can target this thing with your nanocam's built-in incendiary device."

Bruce did so and listened to Leonardo's countdown as he held his breath.

"Three, two, one, launch," Leonardo whispered into Bruce's ear.

Nothing was to be heard or seen by Bruce since the device launched from the camera mounted to his head was too small for the human eye to see. Bruce waited with bated breath. A few moments later, there was an audible popping sound, followed by Leonardo Hoffman saying, "Just wait a minute while AI is re-scan-."

At that moment, a giant explosion knocked Bruce

backward in the air onto his back, and in his panic and disorientation, Bruce jumped up and started sprinting past the flame-filled observatory building. As he looked at the facility, he saw the now exposed telescope inside; it was destroyed. He heard Leo yelling, "Bruce! Don't stop for a second, you get your skinny ass back down here! We have to get out of here, NOW! Nobody inside the building could survive that!" Bruce turned his head back to the direction he was running and sprinted out of the open and back towards the path he had traveled up the mountain.

Once Bruce had his wits about him, he yelled to Leonardo Hoffman, "Leo, you stay right there! I'm coming to you; make sure your nano defense is running, and don't take your eyes off the opening of that cave!" Bruce kept a steady but fast pace down the mountain pathway. His quads were burning furiously from running downhill on the unsteady path, but he could endure the discomfort without worrying that his legs would give out. In just twenty minutes, he ran the forty-five-minute hiking trail and reached the cave opening where he had left Leonardo Hoffman.

Bruce was covered in dirt and sweat and had several

lacerations on his face from the branches of the trailside bushes whipping him on the way down the pathway. His ears were still ringing badly from the explosion.

Bruce was puffing like a steam engine, both from the adrenaline and the physicality of what he had just done, "Leo! Am I glad to see you!" Leonardo picked up and threw a bag at Bruce which Bruce caught, just as he realized he had left his backpack at the mountain's peak after the explosion.

It must have fallen to the ground when the blast knocked me off my feet. Bruce watched Leonardo get down on all fours beside the small pool in the cave, reach down, and grab the container of supplies, pulling them up with a concerned look.

"Smart Leo. Do you have everything?" Bruce puffed.

"Yes, Bruce, it's all in the bag and this container."

Leonardo charged up to Bruce, thrust his arms around Bruce, and hugged him tightly.

Bruce patted Leonardo on the back and said, "Okay, man, now let's get out of here."

Leo nodded, and the two companions left the cave area and returned to the main pathway to the old car parked at the start of the hiking trail.

AI scanned the car and surroundings and determined that there were no devices on or around the vehicle, so the two men jumped in the car, and it sped off up the highway towards Tucson.

After Leonardo had placed AI back on the dash, he pointed back behind him, "Bruce, home is that way!"

"I know, Leo," Bruce snapped back, "We need to get out of dodge, and a motel or something outside of Tucson might be a good place to lay low until morning. We've got to regroup and devise a plan, and we need a new telescope, quick!"

The old '76 Chevrolet Impala sped down the roadway, containing the two men who were planning and theorizing. All the while, AI was scanning the highway; internally, it was analyzing the afternoon's events. When the two men settled down in their motel room at the 'KO motel' an hour later, and the curtains were adequately drawn, AI gave the holographic presentation of the day's events to the two scientists. They had learned a few things about Shadow that night: some of the technology he possessed and his cold tactical prowess.

As Shadow walked down the deserted street in Willcox, examining house addresses, Shadow received the electronic message:

Target 1 destroyed, nanocam 1 destroyed.

Shadow barely broke stride after seeing the update, and as the houses passed by, the stranger glanced at the street sign, turning onto Fremont Street. The corners of Shadow's mouth turned up in a smirk as the thought came to the killer's mind: *I hope you're going to be more of a challenge than this, Bruce Hayden.*

Chapter 33: Back to the Drawing Board - Again (29.11.1981 18:46 hrs.)

Doctor Leonardo Hoffman sat at the end of the single bed in the motel room with his head in his hands. "Play it again," was his command to AI to repeat the report of the day's events.

Bruce groaned, "Leo, honestly, we've seen it like fifty times already."

Leonardo's response was quick and concise, "Don't care."

Given what had happened, the AI sprang into action and replayed all the parts of the day, including its assessment of Shadow and his capabilities.

Leonardo spoke out loud this time as he watched the images floating around them in the dark motel room, "So, on the drive to the mountain, there was absolutely nothing detected out of the ordinary by AI, no Alpha or Beta signatures from Shadow. We stopped at the side of the road to chat about the care package I sent us, and then we parked at the base of the hiking trail, which, in hindsight, was dumb, given what happened today, but AI detected nothing there.

Then we went to the cave and retrieved the package, and AI detected the Alpha, Beta, and Gamma signatures in the water and the box from its time travel, so we know AI is functioning correctly while scanning for those frequencies." Leonardo rambled on while Bruce repeatedly flipped a KO Motel pencil up in the air and caught it each time.

Leonardo continued, "Then we got you set up with your new kit and weapon-enabled nanocam, and you went up the mountain. AI detected nothing out of the ordinary, neither a footprint of the right size nor a recent footprint. There was that brief stop near the clearing partway up the mountain when there was a skunk, then nothing again up to the summit. Then, when you got to the clearing near the observatory, I called it out the moment AI detected the 8GHz waveband frequency.

As I tried to scramble any potential tracking protocol by that nanocam by pulsating your nanocam's frequency, it started to transmit differently; it was transmitting a command or encrypted message. Then, when your nano weapon locked on and was airborne to destroy the suspect device, a last command signal pulse was sent right before it was destroyed. It obviously knew it would be hit with the projectile and most likely sent the command to whatever explosive device Shadow had planted in the Observatory building.

That explosion was far too large to be from that nanobot; it had to be some other kind of incendiary device.

The question is: does Shadow have that kind of 22nd-century weapon here, or did he use materials from this time? And why didn't AI detect the 8Ghz frequency until you summitted the mountain? That frequency could easily have been transmitting hundreds of kilometers."

Bruce finally interjected the doctor's thought process; he was frustrated at the circular motion of their motel brainstorming session and irritated but not surprised at Leonardo's lack of tactical knowledge.

"Leo, Shadow could have easily made that bomb with materials you can quickly get in any of the towns around here. Plus, the analysis of the explosion that the camera caught and the composition of the smoke the nanocam captured as I was running past the fire, both show that it was a simple chemical reaction inside a fragmentable canister intended to destroy the telescope with shrapnel—a pipe bomb. Also, the infrared detection parts of the telescope have a cooling system that would be combustible under the right conditions.

As far as the 8Ghz nanodevice that set off the bomb, once it detected me and the nano weapon, Shadow could easily set the range to be the top of the mountain right outside the observatory and no further. That way, we wouldn't have detected it until we were physically standing in front of the observatory, and he blew up the telescope with me standing right there. Shadow is tactically brilliant and fast, and I don't think for a second that he was caught off guard that day he shot you; he planned to transport you to this place from your office using your fingerprint and your retinal scan, I bet."

Bruce started flipping his pencil repeatedly as AI responded to his comments and calculated the probabilities of the scenarios he had just hypothesized.

Leonardo continued as if Bruce had never spoken, "So, therefore, he must have brought detonators; he could have easily brought a small canister of half a dozen or more. The canister itself would be tiny. The last transmission from the nanocam before it was destroyed was a command to a detonator inside the building that Shadow brought with him from the future, but that last short-pulsed broadcast had a sweeping range. A notification to Shadow."

The holographic projector showed the pulse-like wave in 3 dimensions as its focused beam traveled to the detonator inside the observatory building. Then, it showed the short-pulsed message sent immediately after communicating to the detonator, which was a long-range signal.

Leonardo didn't skip a beat as he watched the visual, "The command to the detonator was entirely in the 8GHz bandwidth, and there are no devices on Earth in 1981 that would respond to an 8GHz signal to detonate. Right before the camera was destroyed, the pulsed message was sent out in all directions with a calculated range of approximately five hundred kilometers. That doesn't help us narrow down Shadow's location one bit."

Bruce spoke up, this time he didn't bother to stop flipping the pencil up into the air and catching it over and over, "Okay, so Shadow brought nano cams with him; he probably has at least half a dozen more if he plans as we do, and he probably has a bunch more detonators, and he has access to more bomb-making materials that he can get around here." Bruce stopped flipping the pencil and looked at Leonardo, "We have to assume he has all those things." He shook his head, "I still just can't fathom how he got into the HEPA lab with all those things when he attacked you. He wouldn't have had a chance to grab anything else after he ambushed you; he was busy pistol-whipping and shooting you!"

"Ha, ha," Leo said sarcastically. Then he lowered his head and said somberly, "And he killed Gus."

Bruce apologized. "I'm sorry, Leo, I'm just trying to keep this dark situation a little light-hearted. I'm sorry about Gus. I know he was your friend."

Bruce paused momentarily and then continued, "Please tell me there weren't any more of these tech weapons in the HEPA lab that Shadow could have accessed before he assaulted you? You had to have stored the weapons for the package you had sent back for us; how the hell did you even get those into the HEPA lab?" Bruce couldn't help but smile at Leo, who had a brief smug look on his face before it returned to his stressed complexion. "Just tell me that what was in that container for us was *all* the illegal stuff you smuggled into that lab, Leo."

Leonardo was already nodding, "Yes, yes, that was it; there was nothing else in that lab that Shadow could have taken, let alone detonators or weapon-enabled nanobots."

"But if *you* got those things in there, how secure is the HEPA lab, really?" Bruce looked at Leo with a puzzled look on his face. The HEPA lab was touted as being an ultra-secure facility. "Not only did someone get in there undetected, but he had detonators, nanocams, and a pistol with ammunition-."

Leonardo cut Bruce off, "Those were ceramic rounds. We have the fragment from my shoulder back at the house, and as far as the detonator and the accelerant used for the projectile, it has to be a substance that is not detectable by our scanners on the way into CERN and the HEPA lab.

I wonder if it's a magnetic-powered gun, which would explain the quiet discharge without a visible silencer and no chemical accelerant."

Leo thought for a moment. "Now that I think about it, is it possible that Shadow has weaponized nanocams or nanobots, and he just didn't use one for the observatory?" Leonardo looked at Bruce with an acutely panicked look on his face.

Bruce took a deep breath and let it out slowly before he spoke, his cheeks puffing out as he did.

"Leo, we have to assume that Shadow does have weaponized nanobots; he could be putting out a red herring for us. Tactically, Leo Shadow probably wants us to think he doesn't have that technology because he knew we would be analyzing the hell out of this incident. He assumed I would be wearing a nanocam to record it.

The only part he failed was that I wasn't killed in the blast. Imagine it: Shadow could have easily put that nanocam from the bushes inside the observatory so our nanocam wouldn't have detected its electronic signature until I was physically inside. He blew that thing up right in front of me on purpose! He may not have even meant for me to be killed; this guy is a psychopath! It gives me goosebumps, Leo, just how calculating this guy is. It's like he's playing a game with us. He's toying with us."

Leonardo looked confused, "Why would anybody do that?"

"Because he gets off on playing this cat-and-mouse game.

Because he likes to play with his prey before eating it. Because he's a psychopath."

Leonardo raised his hand, "Okay, okay, you made your point.

Let's take a break. You try to get some sleep. We had better have that car back tomorrow morning, and we only have tomorrow to get a concrete plan together before you see Kevin on Tuesday."

Bruce yawned at the thought of some sleep, "What about you, Leo?"

"I am going to finish the plan to find a new telescope, and we've got to get you and Kevin out of dodge and fast. I'm thinking a year's sabbatical will do it. Something I can send on their archaic university e-mail program. I can send the communication to their internal e-mail from the nanocam you left on Kevin's diploma." Leo became immersed in the holographic light working with him on the task.

Bruce yawned again, "Yes, a year's sabbatical will do the trick if we can pull it off."

Chapter 34: Fran's Unintentional Ruse

Fran O'Shea walked off the high-speed train into the busy Edinburgh streets; it no longer felt like home to her, but there was usually a sense of comfort in being in her birth country. But not today. She navigated the crowded streets, absorbed in the sequence of tasks she needed to perform that day when she stopped awkwardly.

She looked at the old sign hanging over the sidewalk that read, 'Tollbooth Tavern.' She realized she was where Leonardo Hoffman, her friend, mentor, and co-worker, had instructed her to go directly upon arrival in Edinburgh.

What the hell am I doing, honestly? Fran thought as she pushed open the tavern door and walked inside.

It was 3:30 on a Tuesday, and very few people were in the tavern. The large, Scottish man behind the old-fashioned bar spoke to her, "Come on in, lassie, have a seat. Can I bring you a pint?"

Fran smiled at the familiar accent; it reminded her of her father. She went to sit down at one of the booth-style tables near the back of the tavern. *Okay, Leo, what now?* Fran thought as she sat down, and the red-haired, red-bearded man came over to see her.

"Aye, a Scottish lass, I can tell," he said as he put the electronic menu on her table.

"Aye," Fran said proudly, "I'll have a pint of the house brew."

It felt good to embrace her native accent as she smiled at the man. He left with a grin and a snap of his fingers while singing, "She'll have a pint of the house brew," as he disappeared behind the bar and then out of sight.

Okay, Fran, keep it together. She repeated the scripted phrases before the waiter returned with her drink. She looked around the tavern; it was a little run-down but not too severely, and she wondered if her parents had ever come here before Geneva, in their lives, before little Franny's schooling uprooted them, and before Fran's destiny had changed. She was such a bright young woman, yet she couldn't fathom how Leonardo Hoffman would give her further instructions, possibly from beyond the grave. *I have to stay positive. I know in my heart that he's alive,* Fran thought as she saw the stout red-bearded man return from behind the bar with her pint of beer.

" 'Ere you go, Lass, why so sad?" The man looked at Fran with understanding eyes; he must have heard many stories from people who had come in to drown their sorrows in a pint of beer over the years. He put the beer in front of her and pointed at the electronic menu in front of her.

"We make a hell of a fish and chip 'ere, even though we aren't particularly fond of the English." He laughed and smiled at her again.

Fran took a deep breath as she was staring at the menu.

She looked up at the broad-shouldered man; *okay, Fran, here you go.* Fran thought before she spoke, "You look just like a cousin of mine from Wigtown," her cheeks flushed. *This is ridiculous, Leo. I'm going to look like a fool here,* Fran thought.

The smile disappeared from the bartender's face, but he recovered quickly, nodding his head, "Yes, Lass, maybe I do. What's your cousin's name?"

Fran gulped her beer before answering. *I'm going to need more beer,* she thought.

*"*Uh, my cousin's name is Seamus. Seamus Regan.*"* Fran stammered.

"Aye, Lass, I'll take that as a compliment; I know Seamus well. I'll be back to take your order in a jiffy."

The bartender looked around the tavern as he walked past the bar and into the back of the pub.

Fran's cheeks still felt flushed; she felt crazy having this nonsensical conversation with this bartender she had never met. *She didn't have any cousin named Seamus Regan.*

Stay calm, Fran. You need to go through this, whatever it is.

As Doctor Fran O'Shea drank her beer, she mused about how Leonardo Hoffman might have had tentacles this far-reaching in the world.

I hope you're alive, Leo, Fran thought as she put her beer down. The thickset Scotsman came out from the back of the tavern, looking around him again as he walked to Fran's table.

"Sorry, Lass, the cook just had to go home sick. I will have to close the tavern until my second fiddle cook gets here. You might want to use the restroom before you leave after guzzling that beer." Fran looked down at her beer. It was almost gone already—*stress drinking. Great,* Fran thought.

The man pointed at the hallway behind Fran and spoke again in a softer but more severe voice, "It's not the first or second door; it's the third door on your left, Lass." The large man turned around and told the other two people in the tavern that they were closing, and he showed them the door in not so many words.

"Aye," Fran mumbled as she got up and walked down the dimly lit hallway. She passed the first door, which said "Lads," then slowed as she walked up to the second door, which read "Lasses."

Fran jumped as the bartender's booming voice

called down the hallway, "Yes, Lass, the third doorway," He emphasized "third" in his heavy Scottish accent. He waited until Fran had passed the second doorway and turned to speak after engaging the deadbolt on the front door, "Nice to meet ya, lass; you'll have to go out the back doorway when you leave."

The large red-haired man left towards the bar out of Fran's sight. Fran reached the third doorway with the label 'All-Gender' and slowly pushed it open. It appeared to be an ordinary all-gender inclusive bathroom, but as Fran walked in and the door swung shut, there was an audible click.

The door locked.

"Oh no!" Fran gasped as she went back to the doorway. She started to panic, so she grabbed the door handle and pulled frantically. She stopped dead and turned around when she heard a familiar voice.

"Hi, Franny." Fran spun around, and in the bathroom air space was Doctor Leonardo Hoffman's holographic image. Fran had no words. She just put her hand on her mouth so she wouldn't cry out as she saw her friend's smiling face. Tears were streaming down her face as she watched him.

The holographic video continued as Leonardo spoke calmly, "If you're seeing this, Fran, firstly, I'm sorry. I'm sorry I've dragged you into this." He paused before continuing. "I trust you've done all the other tasks thus far in the proper order before getting here. I trust you, Fran, more than anybody. You are a true and loyal friend, and those are hard to come by."

His look changed from apologetic to serious before he spoke again, "Since you are most likely being tracked, this room and the hallway leading up to this room have been frequency-scrambled, so no audio or video recordings can be electronically captured, so listen up, kid, because this video will only play once, and then it will be destroyed." Leonardo paused before speaking again; he appeared to be searching for his words, although, as Fran suspected, the Leonardo recording the message was more likely lamenting at the thought of what exactly had happened that Fran would be watching the recording.

As Leonardo continued, Fran wiped the tears from her eyes as she tried to see and listen to his instructions,

"Fran, you need to go and talk face-to-face with the director; tell him you have a message from me. He will likely meet with you in his private office space."

"Yeah, right." Fran said sarcastically, "As if that guy will meet with..." Leonardo's following words cut off Fran.

"You may not believe me, but you've got to trust me. He will meet with you and won't be happy with me. In fact, he will be furious with me. And it is imperative that you bring Maryam from the board of directors with you. Insist on it. I want you to approach Maryam and ...."

Doctor Leonardo Hoffman's holographic ghost spoke for the next ninety seconds in the enclosed, electronically shielded restroom as Fran concentrated on his every word. The information Fran received at the Tollbooth Tavern that afternoon would thrust her like a dagger into the middle of the conflict in a way that would change the course of history.

By the time the holographic doctor Hoffman finished speaking, Leonardo's words from the outside table at the quaint little Swiss café years ago returned to Fran as she wiped the fresh tears off her face. She splashed water on her face from the nearby sink, "You have to understand that your role is the most important in that protocol, and you cannot deviate in any way from the instructions I am going to give you."

Chapter 35: The Sabbatical (30.11.1981 10:38 hrs.)

The phone rang in Doctor Kevin Nault's office several times before he answered it. Kevin was still reeling from the news he had received the day before: *the Steward telescope was destroyed, and two of his colleagues were dead.*

"Hello," Kevin answered as he picked up the black receiver off the telephone base from an awkward standing position over his desk. As always, he started twirling his index finger in the coiled cord.

"Oh, hi, Doctor Nault. It's Bruce." Bruce had just about hung up when Kevin answered the phone.

"Hey Bruce, where are you? I didn't see you in the building this morning." Kevin continued talking without waiting for an answer from Bruce.

"I have some bad news, Bruce; the Steward telescope was destroyed in some kind of chemical explosion yesterday, and two of my colleagues were killed in the blast." Kevin sat in his office chair as he spoke; he still couldn't believe what had happened.

"Uh, yeah, doctor Nault, I'm sorry; I heard the bad news too. Were you close to the two researchers killed in the explosion?"

"No, not really close, you know, but I knew them well enough. They both work out of the Tucson campus of the university, so I peer-reviewed some papers for them, and they did the same for me. It was more of a professional relationship, but still, it was so sad that such a catastrophic accident could happen with the coolant for the telescope. It's just unfortunate, and now, selfishly, I also have no place locally to do my research and I'll be honest, I won't feel safe around the telescopes that have that same coolant system until the investigators can figure out what happened."

"I'm so sorry, doctor Nault," Bruce replied. "I hope the investigators get to the bottom of what happened and I can still start shadowing you when you get some telescope time at another telescope." *Shadowing,* Bruce thought. *I'll have to stop using the word Shadow.* It left a nauseous feeling in his stomach when he realized the link he made to the killer's name. Shadow was still out there, and he was not giving any signs of fighting fairly.

"About that, Bruce," Kevin responded, "it is such a weird coincidence that I received an e-mail communication just this morning of an offer to do a research sabbatical for a year at the brand-new telescope in New Mexico at the Very Large Array. An anonymous benefactor has requested me personally and will pay the entire grant for the year."

"Oh, that's great, doctor Nault," Bruce chimed, "congratulations, are you going to accept the offer?"

There was a long pause. Bruce could tell that Kevin was thinking and weighing his options. The shock of the explosion and the perfect timing of a year of research sabbatical was especially appealing to the scientist. He had never done a full year of research; even as a graduate student, he still had classes to teach and supervision in the labs, and Bruce knew that the offer would likely be irresistible. Leo had to construct it that way. This was a crucial moment for the mission.

Kevin started to speak, "Bruce, leave it with me for the time being, but I would like to accept this opportunity to focus purely on research for a year, but it would mean leaving you behind here to keep working on your undergraduate degree."

Bruce sighed, "Yes, I understand, doctor Nault, but I was thinking just now, what if I came to visit you in New Mexico? What if I took a semester away from school in the new year? We could continue our work on quantum computing and how it could help analyze telescope data. I would certainly pay my way to get there and for accommodations and all that."

Kevin Nault thought for some time before answering, "Well, Bruce, I couldn't ask that of you to take a semester away from your studies so I can selfishly collaborate with you about quantum computing."

"Doctor Nault, this is what I want to do, and you are the one I want to learn from, and you are the one I want to collaborate with about quantum computing. Really, what's one semester going to take away from me? I'll return afterward and be fresh and ready to continue my studies. What do you think?"

"Well, Bruce, I do like the sound of that being incorporated into my research sabbatical, and I want to collaborate with you on this. You did ignite this scientific flame in me regarding quantum computing as it relates to analyzing telescope data."

There was another long pause. Bruce looked at Leo, who was looking at the AI display, keeping up marvelously with the conversation. AI was displaying a probability of 94% that Kevin would accept Bruce's proposal. *Not high enough.* Bruce stared at the holographics and waited for Kevin to answer.

"Doctor Nault?" Bruce asked.

"Yes, Bruce, I'm sorry, I was just thinking. Yes, I would like that. I will personally write you a letter of support for you to take that semester away from school to assist me in New Mexico."

Leonardo silently jumped up joyfully as Bruce answered Kevin, "That's great! You won't regret it, sir."

Doctor Nault's reply was downright prophetic, "I know I won't, Bruce. We will do great things together."

"Yes, sir," was Bruce's reply.

As the conversation ended, AI kept working away on its own, modifying its projections and conducting billions of simulation calculations per second, given the drastically new data it had received in the last few days. Leonardo Hoffman continued to work with AI on a plan to secure telescope time in Hawaii, not the VLA in New Mexico. They needed to ensure that nobody in Arizona knew where Kevin Nault would be for the following year.

The tactical choice of one of the telescopes in Hawaii was twofold: it was an island that Shadow would have to travel to by plane or boat, which would mean taking a more considerable risk as far as being caught by local authorities with a forged passport. The second reason was that even though all the telescopes were on the main island of Hawaii, the Canada-France-Hawaii telescope was a perfect segway to liaise with the Canadian and French telescopes once the initial discovery of NH-1983 had been made in July 1983.

Another bonus reason was that due to the clarity and power of the imagery on the new Canada-France-Hawaii telescope, Bruce and Kevin could make their discovery weeks or months earlier, which would be unexpected to

Shadow. The discovery would come at a much more well-known telescope and involve three countries at the onset, but since Hawaii was a state in the USA, Bruce and Kevin could still address the US Congress to begin the Spaceguard project.

Collaboration with Canadian and French astronomers to confirm the trajectory of NH-1983 would yield the same results as the previous missions; however, Leonardo had hypothesized, and AI had confirmed with high probability, that they would be able to drastically change the location where they would create the landing craft and nuclear jet propulsion units. They had to assume that Shadow had an AI unit of his own, and Bruce and Leo's AI was operating under that assumption as Shadow was planning his next move.

The more uncertainties that Bruce and Leonardo introduced into the equation for the killer to find them increased the area of the world that Shadow would have to be looking for them. AI had postulated that once they got Kevin Nault out of the state, Shadow would have to begin phone calling or physically going to many different telescope locations in many other states and countries in search of them. That is after he had concluded that the VLA in Mexico was a red herring.

The mission's trajectory had changed dramatically in the dark motel room where the two men were temporarily living. Unfortunately, Miss Rosie and the house on Fremont Street would have to be abandoned.

AI confirmed in every conceivable simulation that the two men should not return to the house in Willcox that Bruce called his home. Nothing they left there would be helpful to Shadow, just some of Bruce's clothes and a modified basement space that was non-functional without the AI unit. They left a nanocam in the corner of the room in the basement, opposite the blacked-out window. That nanocam had not registered any movement or any sign that Shadow had breached the basement area of the run-down house.

Leonardo frequently glanced at the video streaming from the basement of the house, staring in disbelief at the chaos Shadow had caused in the short time since he had accidentally piggybacked Shadow across the Hoffman Einstein Rosen bridge into 1981.

The two men traveled back to Willcox from their motel room under cover of the night to install some of their newfound technology and ensure the safety of the most important, unknowing participant in their plan.

Chapter 36: Hacking the 1980's (01.12.1981 03:01 hrs.)

Bruce and Leonardo planned with AI throughout the night after they set up their temporary protective barrier around the unsuspecting Kevin Nault. They got back to the motel, and after driving several kilometers out of their way to ensure they weren't followed, they settled back into their motel room.

They were in a fluid tactical situation, and besides ensuring Kevin Nault's safety, they needed to ensure their own safety. This meant they would have to move frequently and not stay in the same motel or rental for more than a couple of days at a time. Leonardo was perfecting his fake ID documents.

"So, Leo, let's go through this again," Bruce said as he paced back and forth, looking at the AI holographic images of their planned travel to Hawaii in two days.

"Okay," Leonardo stood up from his chair as if he were going to give a student lecture, and he walked the two steps over to a small open area in the room where he could go through the details.

Leonardo stared at the holographic planner illuminating that part of the room before he spoke.

"On Wednesday, that's tomorrow; we have the student council group at the Willcox campus as well as at the Tucson campus distributing flyers and putting up posters related to a lecture that Kevin Nault is giving at the Tucson astronomy lecture hall on Friday at 7:00 p.m. Meanwhile, on Wednesday at 7:00 a.m., I will intercept Doctor Nault before he gets into his vehicle to go to the university and convince him that he needs to come with me because of a threat made against him. And before Kevin knows what is happening, we will go to Los Angeles International Airport to catch our flight to Hawaii."

Bruce interjected, "Now, Leo, you can't mess up the story we are telling Kevin. He couldn't know about the mission in any way; trust me, when I told him during another mission, it was not pretty. What I told you and the board of directors was a toned-down version of my last debrief. He had a total nervous breakdown. He is supremely rooted in his reality, and we can't let on that this is anything other than what we planned."

Bruce looked at Leonardo, "stick to the plan, Leo."

"Yes, yes, Bruce, I'll stick to the story. AI seems convinced this will work, but I'm not as good under pressure as you are, Bruce."

Bruce stood up, walked into Leonardo's lecture space, put his hands on Leonardo's shoulders, and looked down at the significantly shorter scientist, "Leo, you got this. Kevin has no way of checking who you are, and in every other mission here, there was an assassination attempt on Kevin. Hence, the past mission data is in our favor, so Kevin will believe you and go with you." Bruce took his hands off Leo's shoulders and stepped back a half step, "Plus, Leo, you're a compelling speaker. Just translate that into what you need to tell Kevin. Pretend you're giving one of your boring lectures, Leo."

"Yeah, lie my face off," Leonardo rolled his eyes at the little jab, which failed to lighten Leonardo's mood. "I'm a bad liar."

"Leo, just stick to the facts. Kevin Nault's life is in danger, yes?"

"Yes, it is." Leo hung his head, knowing where

Bruce's logic was going.

"Then you are protecting Kevin from an assassination attempt and telling him that those responsible for the deaths of the two other scientists at the Steward telescope are also responsible for the threat against him. *It is* the truth."

"Yes," moaned Leonardo. "It's impersonating an FBI agent, and that will be the hardest part. It's like it's from a bad movie."

Bruce continued his therapy slash motivational talk with Leonardo, "Yes, Leo, that part is a lie, and it will be nerve-racking for you, but you are doing it to save not only Kevin, but we need to get to that new telescope and make the asteroid discovery, and that will save the whole world as we know it." Bruce smiled at the last part, and Leonardo picked up on Bruce's patriotic and overzealous delivery of 'save the whole world as we know it.'

"Fine, Bruce, but you need to be careful staying back on your own; I don't like us taking separate flights to Hawaii. I have to be alone with Kevin for two or three days!"

"Leo, Kevin is a friendly and considerate man. Plus, even if you are posing as an FBI agent, you can still make idle chit-chat about his work while you travel, and you get to meet and talk to the great Kevin Nault! Think about it, Leo, between the two of you, you are like..." Bruce started mimicking, counting on his fingers, "the sixth and seventh smartest people I've ever met."

"Ha!" Leo laughed at the unexpected levity in the conversation.

Leonardo sat down on the edge of the old single bed, and it squeaked as he spoke again, "I just worry about you, Bruce. Staying back by yourself is a risk. What if Shadow does something unexpected again, and I'll have most of the mission tech with me?"

"Well, Leo, as they say, the best defense in an offense."

"I will lead Shadow on a jolly chase, and if I'm lucky, I might bait him into being right where I want him to be as my plane to Hawaii leaves the continent."

Bruce changed the topic. "Leo, let me see your duffle bag again. You practice packing and unpacking the nano equipment, and while you're doing that, practice the lines that you'll be saying to Kevin."

"Another deception!" Leo was exasperated. "I'll try to smuggle all this tech onto a plane through all that security!"

"Relax, Leo, their security systems and detection equipment are limited to hand searching and, at Los Angeles Airport, they will scan your bag with x-rays. Just x-rays, Leo. I think you forget how primitive their technology is here."

"I certainly don't forget where I am, but maybe I am getting worked up for nothing. X-rays will show nothing but a pixel or two of a dark spot in my bag, and they haven't even invented the stuff I'm carrying, so how would they know I shouldn't have it?"

Leo was reasoning and self-talking, and Bruce just let him as he flipped his hand in the air and began examining his part of the mission to move the entire operation to Hawaii with the assistance of AI. As he scanned through the locations and the timeline, AI interrupted Bruce's thoughts as it gave a notification that it had captured incoming relevant news from the local news station, as well as a thermal anomaly coming from the nano cam at Bruce's house on Freemont Street.

AI displayed the graphic from the news station with the headline and information from the broadcast.

Suspected murder, suicide in Willcox, AZ. Police have not released the names of the one male and one female who were found deceased in a quiet Willcox neighborhood late last night.

Police have no suspects at this time and are still investigating the multi-house crime scene where there is an ongoing structure fire.

The AI projection changed to show the reporter's video feed at the Willcox crime scene. AI paused the footage and highlighted the address on one of the houses in the video: Miss Rosie's house. The house next to it in the video was on fire. It was unmistakably Bruce's house. The nanocam inside the basement confirmed the elevated temperature and identified toxic smoke.

"Oh, poor Rosie," Bruce mumbled as he put his hands through his hair and put his head down. Leonardo had stopped what he was doing and saw the information that Bruce had been watching.

"Oh, no. I'm sorry, Bruce," Leo sighed. "I know you liked Miss Rosie. Who do you suppose is the other male? It can't be Kevin." Leo's voice trailed off as he realized the implication of what he just said.

Bruce still had his head down with his hands on his head. He was thinking. With Leo's comment, AI silently sprang into action and verified, via the nano cams that Bruce and Leo had installed earlier that night, that Kevin Nault was in his house safe and sound.

Leonardo went over, stood beside Bruce, and touched Bruce's arm. "Bruce, are you okay?"

Bruce looked up after several moments. "AI, search nanocam on Freemont for any sign of movement or suspicious noise, and graph the basement temperature from 2:00 a.m. onwards." AI completed the analysis in seconds and graphed the inside temperature. It then postulated that the fire had been started from outside the house with an accelerant due to the faint, subsonic 'whoosh' that was detected when the fire had started close to 2:00 a.m.

Bruce had his head back in his hands for some time as he weighed their options.

"Leo, pack your stuff. We have to move on this plan this morning." Bruce said through his hands.

"This morning!" Leo yelled. "It's 3:30, and Kevin leaves for work by 7:00!" Leo stopped in his tracks and looked at the stressed-out man beside him.

Bruce took some long, deep breaths with his hands still on his head and his elbows on his legs. *They had drastically underestimated Shadow's capabilities.* He stood up and started packing some gear while his mind revved about six steps ahead. He stopped briefly to respond to Leo. "Get your stuff packed. We move this morning. And you stay glued to that AI feed of Kevin's house." Bruce put on his ballcap and opened the motel room door. It was drizzling outside, and the dim fluorescent lights outside the motel flickered in the nighttime darkness.

"Where are you going?" Leo shouted and threw up his arms in disbelief as Bruce charged out the door into the morning air.

"To get you a car to pick up Kevin in three and a

half hours, of course!" Bruce didn't even look back as he spoke and disappeared into the early morning darkness outside Tucson.

It would not be hard for Bruce to find an appropriately FBI-style car in the city, nor would it be hard to steal the vehicle, since hacking the security systems in even the most advanced car in 1981 was similar to cutting a bike lock with a pair of bolt cutters. By 1981, most General Motors cars had a CCC (Computer Command Control System), a primitive technology that Bruce's watch could easily communicate with and override any locking system, and fool the car into thinking the key was used to start the engine. The most considerable risk that Bruce was taking that night was that the police would soon be looking for the stolen car, and Leo and Kevin didn't need that kind of attention as they tried to escape the area.

Bruce ran flat out for almost ten minutes when he came upon a more secluded house with a nice boat-looking car that would serve their purpose nicely. He puffed his cheeks out as he settled his heart rate after the burst of physical activity; the run felt good, and the adrenaline of what he was about to do just helped his nervous energy turn into focus.

The house was completely dark, and as Bruce used his watch to interface with the car's onboard computer system, the doors unlocked with a 'clunk,' as Bruce closed the door as quietly as he could, the engine roared to life. Bruce left the headlights off as he quietly backed out onto the highway and then sped off to the motel to pick up Leonardo.

On the way, Bruce stopped at a nearby residential parking lot, got out, expertly unscrewed the rear license plate from a similar-looking car, and was on his way. All the exhilaration from the two thefts he had just committed was funneled into the one thing he felt at that moment: determination. The years and years of training served their purpose as Bruce planned the hours to come: the extraction of Doctor Kevin Nault from the closing grasp of Shadow.

Chapter 37: Fran and Maryam Meet with the Director

"That damn Frenchman will never see the light of day again once I get my hands on him!" the director screamed as he turned away from Fran and Maryam, mainly to compose himself after what they had to say. *How did that little coward do all that sneaking around under my nose?* the director thought as he nervously stroked his goatee. A bead of sweat had run down his forehead and was expertly perched at the end of his nose as he thought.

Fran and Maryam looked at each other nervously; the director had virtually unchecked power, and he was fuming mad, as Leonardo had predicted. Fran couldn't help but smirk ever so slightly at the genius of Doctor Leonardo Hoffman, but that smirk was transient as the director spun around in his chair from his many projection screens and silently looked at both women intently; his piercing gaze would have crushed most, if not all, of the board of directors, and definitely Mitch.

I wonder what that wimp is up to while I stand here taking this heat, Fran thought as the director looked her up and down. Fran and Maryam did not flinch in their resolve, nor did they back down in their posture as the director thought of what he would do next.

Leonardo made an excellent choice, teaming up Maryam with Fran to deliver the message to the director; Maryam was as stubborn and confident in her determination as Fran.

The director's facial expressions changed from angry to intense study to calm confidence right before their eyes. His calm poise gave the air of some epiphany, and when he spoke again in his regular deep, monotone voice, it gave Fran chills, reminding her of her conversation with the director on the phone after the lab incident.

"Ladies," he opened his hands to the side of his chair in an embracing pose. "What would it take for me to convince you that Leonardo Hoffman has lied to both of you? Can't you see, for some reason, *he* has sabotaged this mission? Maybe the stress of his position has marred his judgment. Maybe he is self-sabotaging because he can't stand the thought of another mission failure."

Fran had already taken in her breath to defend Leonardo Hoffman after the director's first question, and when he finished speaking, she let it out, Scottish style.

"How dare you question Leonardo's integrity. He is the reason we are standing here talking about this mission. He's the reason we have the technology to complete this mission, and he's the reason that this mission will still be successful. You sit down here like a weasel in its den and scheme and plot and pull strings where you think nobody can see you." Fran's face flushed as she saw the director's face turn redder and redder until she thought it might just pop off.

She stopped momentarily as her outburst echoed in the underground space, and she lessened her tone considerably before speaking again.

"Mr. Director, I have no doubt whatsoever of Leo's integrity nor his resolve to complete this mission."

The director smirked condescendingly. "And how do you know, little Franny, that your friend is still alive?"

"Don't call me that," Fran growled. She was inhaling to let the director have it again, but she caught herself, and after a glance at Maryam, she let her exasperation out quietly this time. *This had to be diplomatic. 'Don't fall for his bait to get a rise out of you. You now hold all the chips in this game,'* Fran remembered Leonardo's words from the holographic message in the pub restroom.

Fran composed herself. "Mr. Director, I don't know for sure that Leonardo is alive, but what I do know is that I will see this through because I believe that Leonardo has ensured the success of this mission despite your mishandling of the assault on the lab, and the theft of the twin HERB device from Leonardo's desk."

The director shot back, "Fran, think about it: your friend goes missing, and at the same time, the second HERB device goes missing. All without a trace. Let me tell you what I think happened." The director leaned in and spoke more softly but still condescending, "I think Leonardo was under a lot of pressure because he felt like he was losing his *perceived* control of the mission. So much so that he staged that stunt in the lab, with the cooperation of Gus Maxwell, and they somehow both left using the one HERB device."

Fran's pulse was galloping. "What do you mean, they both left using the one HERB device? Gus is dead. I saw him lying there. The police put his body in a bag on the stretcher that day."

The director gave a cold smile, "Did they?"

Fran's heart was raging inside her chest. *Could Gus still be alive? Impossible; I saw him lying there in a pool of blood.*

Maryam, who always seemed to be calm and relaxed, took over for the sputtering Fran and spoke in her confident and fluent European-English accented cadence, "Mr. Director, if that's what happened during the assault on the laboratory, why would you hide it from your report to the board of directors? If you recall, I was there when you presented the findings, and you said nothing to imply what you're saying now. What proof do you have besides your belief? I think you're trying to snow us. It won't work."

The director shifted in his chair before he spoke. *He's uncomfortable,* thought Maryam. *He's lying.*

The director spoke as if he sensed what Maryam was thinking, "Maryam, what purpose would it have served to give that conclusion of mine to the entire board of directors when we all last met? It would have resulted in nothing but chaos, and most would have lost hope in the mission, and we'd be in the same position we are now." He extended his arms again and spoke grandiosely, "*We* would be here, talking like we are now." The director's voice echoed inside the underground chamber.

Fran was living up to her nickname, the bulldog, as she charged at the director again, "Mr. Director, please enlighten me; what proof can you show me that will get me to believe that Gus didn't die that day? I saw him lying in a pool of blood. I saw him being wheeled out in the body bag that day."

The director smiled; he was enjoying this. He paused a moment before he spoke. "Fran, did you *actually* see Gus' body? Or did you see only his shoes and his legs?" The director's arrogance was palpable.

"I saw that he was on the floor behind his desk. There was so much blood around his legs," Fran replied, her voice shaking, "are you saying that it was a ruse? Was that staged? That wasn't Gus' body they took out of the HEPA lab that day?" Fran could feel a tear welling up as her anger and disbelief swirled.

"Oh, Fran," the director smirked as he called up a file from his AI unit.

The screens and the holographic projector behind the director blacked out momentarily. Then, it projected a photo that appeared to have been taken by the Forensic Drone, which the police used to document the crime scene at the HEPA lab. The picture was taken from above doctor Gus Maxwell's desk, pointing down towards the floor. Gus's chair was off to the side of the image, against the wall, and there, on the floor, were two disembodied legs with the same pants material as Gus' pants that day. The legs went down to the oversized brown loafers, and there was a significant pool of blood.

"I don't understand; you doctored that photo!" Fran yelled as she studied it. Her confused mind spun around like a centrifuge.

"Did I?" the director grinned. With a swipe of his hands, he called up the next photo, a close-up of the top of the legs and the forensic marker # 58 sitting in the picture just outside the pool of blood.

There was silence in the room. The director then swiped again to reveal an electronic forensic report. As Fran and Maryam skimmed the information, the director highlighted the last few sentences at the bottom. It read,

Blood was determined to be type A negative, and DNA sequencing confirms the blood source is Doctor Gus Maxwell. A 3D printer unit of unknown origin most likely created the fake lower leg segments.

The director swiped once again as he watched Fran and Maryam's contorted, confused faces. The last image from the drone in the sequence was a video that started to play with no sound. It showed the two police officers putting the two fake legs in the body bag where the legs would usually be, inflating a body pillow-shaped object before zipping up the body bag. The director stared at the two women and let the euphoric feeling of complete control linger before he spoke again.

"Ladies," the director started as the images darkened and the lights in the room brightened, "What do you think now?"

Maryam spoke first, "You planted those two police

officers?"

"No, I didn't, Maryam, but in the interest of security for the mission, people need to believe that Gus is dead. Yes, I told those forensic officers to use a buoyancy bag and act like it was Gus' body when they were leaving. I can't tell you any more than that right now."

Maryam rebutted, "But, director, you are leaving out the part that Leonardo has put into play some checks and balances that have caught you in a lie, sir. It's more than just helping to fake Gus' death. What do you have to gain by lying to the board of directors, the HEPA lab scientists, and all the many investors you deal with to get all our funding?" Maryam paused. "And why would you do that when the fate of the entire planet is at stake?"

The director's face was beet red, "I have spent almost my entire life in service to this planet, Maryam. And now, as we approach the climax of this saga, I work around the clock to ensure the success of this mission. Don't pretend that I am sitting where I am, and you are standing where you are because of chance. It is because I am the one who pulls the strings, and I am the one who has dedicated his entire life to this mission. You think that because you come to a couple of meetings per month, you are somehow integral to this mission. I have some news for you, Maryam; you and the rest of the board of directors and your precious Leonardo Hoffman are all glorified approval robots. I come and tell you what we need to do, and invariably, you all vote to give me what I need for this mission."

The director's face became an even darker shade of red before he roared, "And the two of you coming here, to my office, to my home, and attempting to blackmail me, it's pathetic!"

There were moments of silence as the director's thunderous words echoed again in the underground quarters before Fran and Maryam turned in unison and started to leave.

"I'm not finished with you two!" The director yelled.

"Well, we're finished with you," Maryam replied as the two ladies started to leave the room and headed to the elevator to return to the conference room. "Our statement still stands. You'd best think long and hard about it," Maryam finished speaking as the elevator door was closing.

"You're bluffing! He's bluffing!" screamed the director.

The elevator door closed, cutting off the sound of the director's rage. The two ladies traveled up to the empty conference room and eventually out into the fresh Swiss nighttime air, all in utter silence.

Maryam broke the silence after inhaling and exhaling the cool, crisp air. She smiled a nervous smile.

"So, shall we head for a spot of tea?"

As they walked, Leonardo Hoffman's voice haunted Fran's flawless memory.

"Fran, you must assume you will be under constant surveillance once you and Maryam have spoken to the director. You will both have to trust that what you have helped put into motion will propel itself, and it is imperative that the director is bored and frustrated out of his mind when you act as if nothing has happened. You don't even hint or whisper about what you know after you leave that office."

Oh, Leo, I hope you're alive so I can hug you or slap you in the face. Or both, thought Fran as she fiddled with the tiny electronic device in her pocket that Leonardo had left her at the little Scottish pub several days before. The two ladies walked down the well-lit pathways away from CERN in silence.

Chapter 38: The Extraction (01.12.1981 06:59 hrs.)

It was 6:59 a.m. when the black 1977 Buick LeSabre rounded the corner and abruptly stopped on the road at the end of Doctor Kevin Nault's driveway. The driver sat momentarily, composing himself as he watched the subject walk out of his house, briefcase in hand.

Kevin watched the black car from his small cement porch for a moment as he locked his front door, and he continued out of the house, down his front steps, and into his driveway. As he set down his briefcase on the hood of his car to talk to the car's driver, now blocking his driveway, the driver got out and walked over to Kevin with an outstretched hand.

"Detective Jones, FBI," he said assertively as he vigorously shook Kevin Nault's hand. He flashed a badge identification before putting it back in his pocket.

Before Kevin could answer, the man spoke again, "Doctor Kevin Nault, the FBI has reason to believe that you are in immediate danger, and you must come with me for your safety."

Kevin was taken aback and chuckled a nervous laugh, "What? Me? Why would the FBI have anything to do with me?"

Leonardo repeated, "Sir, we believe that you are in danger. I can't explain now, but you must come with me to a safer location to talk. Please, sir, this way."

Kevin Nault looked Leonardo up and down. He stopped and looked at the injured side of Leonardo's face. "I think you've mixed me up with someone."

"No, sir, you are Doctor Kevin Nault, 38 5th Street, Willcox, who graduated from Cambridge University in April 1979 and is currently an associate professor at the Willcox campus of Tucson University. Two of your colleagues were killed last week in an explosion at the Steward Observatory, where you also conduct scientific research, and the FBI has reason to believe that those two men were murdered and that your safety is now at risk."

"What?!" Kevin felt dizzy, so he bent down and put his long arms on his knees to regain his balance. "This can't be happening…"

"Sir," Leonardo continued in his official voice, "Sir, I need you to come have a seat so you don't pass out." Leo conveniently led Kevin to the passenger seat of the Buick and opened the passenger door so that Kevin could sit down.

"I…I…I should hear what you have to say," Doctor Nault stammered to the posing FBI agent as he sat in the front passenger seat of the idling black car.

"Great, watch your feet, sir," rushed the agent. Leonardo started to close the door, and Kevin pulled his feet inside before the big, heavy door closed. Leo ran and grabbed Kevin's briefcase off the hood of the other car and straight back to get into the driver's seat of the black Buick; he tossed the briefcase into the back seat and put the car in gear as it roared into life as it sped off down the street and around the corner.

"Hey! I said I'd talk to you; I never agreed to come with you," shouted Kevin.

"Please, sir, just let me protect you. It's my job. All will be explained to you, sir, but we have a flight to catch at LAX in about 11 hours."

"A flight! Where am I going? I don't have any clothes with me or anything!" Kevin was still panicking but also indignant at what was happening.

"Sir, please calm down. Might I suggest taking deep breaths? Then we can talk about the details. You are safe now. You must trust me for a little longer until I can explain. Your brain won't properly work when you are anxious, and you won't grasp what I'm telling you unless you calm down. Sir."

Kevin Nault complied and took deep breaths as the Buick sped up the highway. For several minutes, Leonardo tried to pretend that he was not looking out of the corner of his eye at the famous Kevin Nault and that he wasn't thrilled to be able to spend time with the scientist that he had come to idolize. He started to speak in short, slow, and calm-sounding sentences.

"First of all, Kevin, can I call you Kevin, doctor?" Leonardo asked.

Kevin's head was leaning back on the seat's headrest, and his eyes were still closed when he answered, "Yes, please just call me Kevin. If we are doing this, what's your first name?" Kevin raised his head from the headrest and looked at the posing agent in the driver's seat.

"Stan. Call me Stan," replied Leonardo.

"Okay, Stan, can we start with where are we going?" Kevin was still looking over, studying Leonardo.

"We are going to the Los Angeles airport. We have a flight in about 11 hours, which will take us to Hawaii," Leonardo replied.

"Hawaii! Why the hell are we going there, may I ask?" Kevin's voice was raising in volume again.

"Breathe, breathe, please doctor… er… Kevin," Leonardo assured Kevin before he answered.

"We are going to Hawaii because we believe there may be an assassination attempt on you, sir, and the Canada-France-Hawaii telescope has all the resources you need to research. It is much more capable for you to conduct your research than the VLA in Mexico, where you had planned to go."

"How do you know about that, and why is my research even important to the FBI?" Kevin was staring at

Leonardo quizzically. *This was like a bad dream.*

"Well, sir, please understand that we needed to get you out of the state because of the assassination threat...." Kevin Nault cut off Leo.

"There's that word again...assassination. You must be important to be 'assassinated,' and I'm just a research scientist. Why are you using that word?"

"You're right," Leonardo responded, "I'm just used to using that word... in my line of work, that is." Leonardo's throat felt dry, but he did his best to stay calm.

There was a short pause before Kevin said, "The FBI set up that sabbatical at the VLA? That is so weird. I still don't understand why the FBI is interested in my research in protecting me or why my two colleagues were murdered. So, you say."

"Yes, sir, I did say that." Leonardo felt better telling the truth and continued, "Those two scientists were murdered, and we believe there is a tangible threat to your safety. Terrorist bombings are the FBI's mandate, Kevin, and we need to get you out of dodge until we can catch the persons responsible for the bombing."

"Bombing?" Kevin's voice rose in intensity. "Bombing! The news said that the explosion was caused by the telescope's cooling system overheating!"

"Sorry, no, that's not what happened. Often, the FBI has to tell the public something incorrect while we conduct our investigation and try to catch those responsible."

"And..." Kevin was gawking at Leonardo.

"And what?" Leonardo asked as he kept his eyes on the road.

"And who did this? Who is responsible for me being whisked away with nothing but my briefcase? And what do terrorists have to do with a few astrophysicists doing, realistically, mundane experiments at a small, lesser-known university?" Kevin was not getting the answers he wanted and was starting to get mad.

"I understand what you're saying, Kevin, and I understand your frustration, but I can't tell you why because we don't know why...yet."

Leonardo's mouth was parched again. He stared at the road and tried to keep his composure. *How did Bruce talk me into this, anyway?* Leonardo thought. He continued, trying to calm the scientist down, "Kevin, these are the facts I can tell you:

Your life is in danger, and the safest place you can be is with me. That means traveling with me to Hawaii. Today.

Your student who has been meeting with you on Saturday mornings, er - his name escapes me right now," Leo lied. "Bruce.

Bruce Hayden is his name. He is also in the same kind of danger as you are."

Leonardo could hear Kevin gasp at the realization that Bruce had gotten dragged into this. "Oh, my God, how could a first-year student be involved in any of this?" Kevin was flabbergasted and started breathing deeply again with his eyes closed.

"The boy's involvement is not your fault," Leonardo assured. *That's the truth,* he thought. "The boy had somehow gotten the attention of the terrorist group when they were doing surveillance on you. We have intercepted some of their communications. It's unfortunate, but the boy will have to come to Hawaii as well until we can get this situation under control."

All Leonardo could hear was the forced in and out slow breathing from the passenger seat.

Kevin finally spoke after several moments, raising his head, "Where is Bruce now?"

"He's with another agent and will be traveling to Hawaii. He will be on the same flight as us, but..." Leonardo stopped talking as he saw Kevin's head back on the headrest again with his eyes closed. "Kevin, I need you to focus on what I have to tell you."

Kevin Nault's head came off the headrest and looked at Leonardo. He looked like a bewildered man whose reality had changed drastically in the last ten minutes.

"Kevin, if or when you see Bruce at the airport or on the flight, you must pretend you don't know him, and you'll have to trust me on that until we get you to the safe house on the island. Don't acknowledge him, don't smile at him, pretend he doesn't exist. Do you understand?"

"Yes, I understand." Kevin was still in disbelief that Bruce was somehow a target. He was shaking his head as he looked out the window and tried to digest the firehose of information that he was getting from the FBI agent. After a few minutes had passed, Kevin said, "So, you've got two facts in your list, Stan, what else do you have?"

"I like you already, Kevin. You're paying attention. They told me you were a genius, but I've never met a genius before." The sarcasm was evident, and Kevin let out a laugh as well.

Leonardo continued, "You and the boy will live at the FBI safe house while you conduct your research at the CFH telescope. Your time on the telescope has been secured, and your whereabouts will only be traceable once we allow it. It may be days, weeks, or months.

You must go through us when you need to communicate with the university about your funding or any other research-related communications. Everyone else must believe that you are at the VLA in Mexico.

And lastly, and most importantly, your safety and the boy's safety depend entirely on your location being kept a secret. Period."

Kevin interrupted Leonardo, "Number four and

number five are pretty much the same, Stan; your list of facts has turned into a list of orders."

"You're right, Kevin. But at least you are paying attention now,"

Leonardo reached over and put his hand on Kevin's shoulder. "You'll be okay if you stick with what I tell you. Okay?"

Leonardo's eyes and voice had softened as he looked at Kevin. "Okay?"

"Okay, Stan," Kevin answered. He had calmed down considerably and still had many questions, but he was a good judge of character, and there was something about the agent that made Kevin feel comfortable.

"What's with the scar on the side of your face?" Kevin asked.

Leonardo put his hand up to the side of his face and felt the scar before he spoke. *It still hurt.* "Uh, I got into a little tangle several weeks ago before I was assigned this case." Leonardo tried to deflect the tension with some humor, "You should see the other guy!"

Kevin chuckled but still looked at Leonardo, "Man, I don't know how you guys do it. Thank you for helping me."

Leonardo could only nod because Kevin's sudden compassion produced a lump in his throat that would have made his voice crack if he had spoken. The two men drove much of the rest of the way to Los Angeles in silence.

Chapter 39: The Mechanics of Moving a Monster

By the 2020s, scientists proved that they could change the orbit of an asteroid after slamming a high-speed object into it. The fridge-sized Dart spacecraft of September 2022 smashed into a one-hundred-meter asteroid called Dimorphos. NASA subsequently proved that the orbit of Dimorphos around its parent asteroid, Didymos, had changed by more than ten minutes.

It was the first time humanity tested the science of moving earth-bound asteroids and the first time that humans had successfully changed the orbit of a celestial object with the goal of future asteroid deflection from Earth. At the time, the scientific community hailed the achievement as a leap forward to protect the Earth.

The wrinkle with the mechanics of crashing a person-made object like the Dart spacecraft into an asteroid was that if the asteroid had significantly higher mass or density, the entity wouldn't have enough energy to change the asteroid's orbit. Slamming the Dart spacecraft into SG-

2131 would be like firing a mosquito at high speed into a moving car; it wouldn't do much except leave a smear on the windshield of the asteroid.

A constant force over time was the only way to move the colossus. In the 2050s, NASA tested the technique to exert pressure on an asteroid to change its course around the Sun. The experiments concluded that diverting a giant Earth orbit-crossing asteroid was possible. Scientists in the early 1900s established the mechanics and mathematics of exerting force on a celestial object; however, the technology to produce such a force over a long period wasn't a reality until the 2060s with the long-sought holy grail of fusion energy.

At the time, a relatively small but constant force on a large moving object in space, courtesy of an array of fusion-powered jets, proved to be a success for humankind. All humanity needed was time.

In 2131, when scientists discovered the killer asteroid and showed it would strike the Earth, humanity had the technology and the knowledge to move the colossus; they just lacked the time. By combining time travel technology with asteroid deflection technology, the human race had a good chance of saving itself. That is if humanity didn't sabotage itself first.

Chapter 40: Getting out of dodge (01.12.1981 16:31 hrs.)

Leonardo and Kevin were standing quietly in a long lineup, waiting to go through security at LAX. Leo nervously watched the crowd around him as he and Kevin inched closer and closer to the security scanners. *If Shadow gets here before we leave, we're screwed,* thought Leo as his eyes darted from person to person.

Kevin interrupted Leonardo's racing thoughts, "Hey Stan! You gotta dial it down a few notches. You're making me nervous."

Leo stopped scanning the crowd, looked at Kevin, and relaxed his shoulders—no *need to stress him out. I can't do anything anyway. I'll have to trust AI to protect us if Shadow shows up.*

Leo stepped beside Kevin and smiled, "Good idea, Kevin."

The two men stood in the ever-shortening line in silence. Leo, with his head down, was planning as they came closer and closer to airport security. He gripped his briefcase tightly in his left hand, and with his palms sweating profusely, he had to adjust his grip on the briefcase often. Leo kept an eye on Kevin, who was also visibly stressed but less than he would have guessed under the circumstances.

Leonardo thought about the methods used for security in his own time: scanning in all types of frequencies, facial recognition, and lip reading extrapolation.

This is not 2131; this is 1981. They can't detect the stuff I have in this briefcase if they haven't invented it yet. Leonardo's self-talk was not diminishing his anxiety. After being thrust into it, he was beginning to have an even greater appreciation for Bruce's mission.

He didn't raise his head as he spoke and kept staring at his briefcase, "Ahem. Kevin, remember, if we are asked, I'm your friend coming to help you set up your new place at the research facility."

"Yes, I remember Stan. You told me on the way here," Was Kevin's half-mumbled reply.

"Sorry, yes, I tend to do that." Leo changed the topic as he glanced up to see how close they were to the front of the line, "So, Kevin, what kinds of things do we need to set up when we get to Hawaii?"

Kevin thought about the question for a moment before answering. "Not a terrible number of things. Whenever I've been invited to research out of town, they provide me with accommodations, and I need to unpack and get some food for the fridge. Then again, you will probably have some things we need to do when we get there, right, Stan?"

"Yes, Sir," Leo replied, "just remember, I'll take care of everything else and brief you on what you need to know. You focus on your research."

"Yeah, I'll try," Kevin retorted as he rolled his eyes at the ridiculous situation he found himself in. "I would love to know what you know, Stan, because I can't fathom what focusing on my research has to do with any of this."

Leo shuffled his feet as he responded, "Well, I just meant you can focus on what you're meant to be doing as far as research, and I'll focus on the part that brings me into the picture, Kevin. We'll work together, and things will turn out fine, you'll see."

Kevin, a keen observer, waited a few moments before he mumbled his reply. He couldn't help but smile at the absurd circumstances,

"Well, at least now I know what you look like when you're lying."

Leonardo Hoffman didn't have time to respond; they were at the front of the line at security.

"Tickets, wallets, belts in the bins please! One at a time, please!" The hulking airport security man bellowed past the two men as they reached the front of the line. Leo took a slow, deep breath and smiled at the security guard as he let out the breath, and their eyes met.

"Sir, your briefcase goes on the conveyor belt to be scanned, please." The security guard motioned to Leo with a sweeping gesture that Leo was sure he had done thousands of times. Leonardo put the briefcase, his wallet, and his belt in the tray and onto the conveyor belt as he stepped towards the second security guard in his lane, who was wielding an archaic wand-looking metal detector. Leo glanced over at Kevin, who had been pulled over to a second aisle and had put his briefcase and the contents of his pockets onto the conveyor belt.

So far, so good, Leo thought, smiling another big smile at the second, even larger security guard. Leonardo stood for the metal scanner to confirm for the guard that Leonardo had no metal on his person and was relieved to hear the man usher Leo along, "Okay, sir, keep moving, please. You can go get your items from the x-ray scanner."

Leo obediently walked over to the conveyor belt just as he saw his familiar brown briefcase coming out of the scanner. He stopped breathing as the guard watching the screen stopped the conveyor. He went over, grabbed Leo's briefcase, and brought it to a metal table.

The security guard looked at Leo. "This is your briefcase?"

"Y-yes, it is, sir." Leo stammered. He stopped breathing and watched in horror as the guard opened the briefcase and looked inside. He looked at Leo again and said, "Did you pack this suitcase yourself, sir?"

"Yes, of course." Leo didn't have to lie about that

part. He shuffled back and forth subtly, trying to understand what the guard spotted inside the briefcase.

The guard's dark complexion made the wrinkles on his face more visible; Leonardo could tell he was looking at something and had put his hands inside to move objects to the side to get a look. Leo's heart raced as he waited. He could see Kevin had cleared security already and was waiting on the other side.

Leonardo's curiosity got the best of him. "Sir, can I ask what's wrong with my briefcase?"

The security guard spun the briefcase around; it was still open.

"It's nothing, sir. I get a little suspicious when I see a briefcase packed so neatly." His expression softened, "Have a good flight, sir."

Leo laughed as he let out his breath and closed his briefcase. "Thank you, sir," he responded as his voice cracked. The security guard didn't even hear Leonardo as he returned to his seat at the X-ray screen, and the conveyor belt creaked back into motion. Leonardo took his briefcase off the table and entered the airport's secure area through the gate. He walked over to Kevin and smirked as they started to walk together towards their gate.

Leonardo was ecstatic that he and Kevin had both gotten through security without any trouble, and he refocused on the plan he had committed to memory. The little dingy motel room he and Bruce had been in hours earlier came to his mind as he mentally went through the steps Bruce had given him.

I hope you're okay, Bruce, Leo thought.

Leo broke the long minutes of silence as they walked with the flow of people, like cattle, on the way to their gates, "Kevin, we are going to walk past our gate and sit at the gate for New York, which is two gates down. It's busier down there, and we need to watch our surroundings even in this more secure area."

Leo scanned around them to see how close people were to their conversation, and he continued but in a quieter tone, "We won't see your research companion at all while we are waiting, and you might not even see him on the plane. You'll have to trust me that he will be okay, too."

"You're the boss, Stan," Kevin immediately replied.

He seemed to be just going with the flow now, and Leo wasn't sure if that was a good thing. Was Kevin a person who felt like reality had left him and was going along with whatever happened? Or was he genuinely dealing with this well? Leo had read in mission reports and been told by Bruce just what a remarkable person Kevin Nault was, but he was marveling at how he had come to be walking with the legendary Kevin Nault. Within a few hours, he would fly to Hawaii with him and eventually witness the great man make the historic discovery of NH-1983.

Fingers crossed, Bruce, Leonardo thought as he and Kevin reached 'Gate 82 – New York.'

Chapter 41: Hunting a Scientist

Shadow had waited for hours near the crime scene, hoping that Bruce Hayden would return to the house he had called home in this town. Not being phased at all by the lack of success, Shadow left the area on foot and walked several blocks, picking up multiple nano cameras, before getting into the beige four-door car stolen from a ratty neighborhood in Tucson.

A short time later, Shadow stopped at the side of the roadway and turned off the car. Folded maps and mission documents were examined closely. The 22nd-century watch on Shadow's wrist verified from other strategically placed nano cameras that Bruce had not returned to the university, nor had he been to the grocery store nearby.

Where have you gone, Bruce Hayden? I've just gotten started with you.

Shadow put the car into gear and got back onto the highway, and as the evening traffic became more and more sparse, Shadow was still driving and thinking.

The stars shimmered over the Arizona desert landscape that night as SG-2131, perfectly cloaked in darkness, hurtled towards the Earth in its penultimate orbit. The asteroid, Bruce, Leonardo, Kevin, and Shadow, their destinies, were now intertwined in a cosmic dance that could obliterate the world.

Chapter 42: Leonardo and Kevin's Flight to Hawaii

I hope you got on the flight okay, Bruce, Leonardo thought as the 747 aircraft boomed off the runway at LAX, destined for Honolulu International Airport. The cigarette smoke was oozing continually through the curtain from the smoking section of the plane, and Leonardo rolled his eyes at the sheer stupidity of tobacco smoking in the 1980s. Leonardo was dreading the long flight, having not gotten a single breath without any toxic nicotine since they arrived at LAX.

I'll probably land in Hawaii with emphysema from all this second-hand smoke, he thought as he glanced over at Kevin, who had put his head back on his seat and had closed his eyes already.

Sheesh! Doesn't this guy get stressed about anything? Leo thought. *I'm worried about Bruce, Kevin, the asteroid, getting to the telescope in one piece, and keeping things together in my head, and this guy is nodding off!*

Kevin opened his eyes and glanced at Leo as if he heard Leonardo's thoughts. Kevin gave a quick smirk when his eyes met Leo's, and then he put his head back again and closed his eyes.

I need to organize my thoughts to land with a clear plan in my head, Leo thought.

Doctor Leonardo Hoffman closed his eyes and started to access his happy place - *his mind.* The plane was a cigarette polluted environment, but Leonardo concentratedly slowed his breathing as he had done thousands of times before, and he brought all the plan elements into his mind's eye, then filed them where they needed to be.

Having a freakishly good memory, Leonardo could visualize elements of a plan and literally go back and revisit conversations, planning meetings, and even short discussions with people and watch them again with almost perfect accuracy. Over the next couple of hours of the six-hour flight, Leonardo mentally examined all his recent conversations with Bruce and inventoried all of the tech in his briefcase that was leaning on his feet under the seat in front of him.

Feeling better and more relaxed after refreshing Bruce's instructions and the plan to make the island safe, Leonardo let his mind wander a little bit, and Fran O'Shea came into view. Leonardo recalled his embarrassing conversation with Fran, Mitch, and Gus on that fateful day in the lab when Shadow ambushed him. Then he revisited the videos he made explicitly for Fran should the mission be compromised, which it had. Wandering further as Leonardo's body was demanding sleep, his mind took him to a place that he didn't expect:

Sitting at the large boardroom table near the HEPA lab, Leonardo met with the director and the rest of the board. It must have been in the debriefing and planning stages for the first mission because Maryam was speaking about initial mission policies, information sharing, and critical policies. The months of endless work to get the HERB mission off the ground had taken a toll on Leonardo. He was looking at his exhausted reflection on the surface of the glass boardroom table, but he was listening intently to all the conversation happening in the room.

"Item 14.5.12, Maryam continued, is related to information sharing and the SafeNet procedures for mission debriefing. Should Bruce be unsuccessful, there needs to be an extremely tight circle of trust related to the information that Bruce provides in his debrief. In that event, we must ensure that only people critical to the mission are privy to the sensitive information Bruce uploads from his AI unit; so that means, in my opinion, doctor Leonardo Hoffman should be the only person from this committee who should have access to that sensitive information from Bruce and his AI in preparation for a possible second mission attempt with another mission specialist."

Leonardo perked up and raised his head to look at Maryam as she was speaking, and he heard the director start to talk in his slow, bass voice. "Maryam, I think I understand why you would suggest that Leonardo be the only one who has access to that information, but I would argue that I need access to that information as well. Not just to have a second observer and a second set of eyes on that information, but what if something happened to Leonardo? We need multiple fail-safes in multiple layers for a mission like this."

After speaking, the director sat back confidently in his chair and watched to see what the other board members would say.

Leonardo could hear the director's heavy breathing from his right as he watched each of the board members speak, and seeing these events in hindsight felt different now. Board members made a few comments, some taking Maryam's side, some sharing the director's view, and as the table came full circle back to Maryam, she looked at Leonardo for a moment. Leo could see now she was looking at him intently, but why? After several moments of studying Leonardo, Maryam spoke, "Leo, what is your position on this?"

Leonardo listened to himself as he answered Maryam. "Well, since fanaticism, sabotage, and espionage are constant concerns for this mission, having a second person in the know as far as mission-critical data is prudent."

Leonardo could hear the director shift in his seat; he was figuring he would get his way. Leonardo continued, "However, I feel that since Bruce will have an independent recollection of the events having lived them, and the AI will have a record of precise locations and dates and times, that we leave the information we don't directly need in Bruce's head, and we do a secondary encryption on the AI unit upon his return. That way, should Bruce be unsuccessful this time, either he goes again with his knowledge and his own AI unit, or he briefs the new candidate directly, and we can essentially copy his AI unit information to the new candidate's AI."

Leonardo could feel his frustration and exhaustion, and he watched as he paused before speaking again.

"Should Bruce be unsuccessful, the mission data from Bruce and that AI unit are critical for analysis to ensure the mission will be successful the second time. I don't see any benefit in anybody at this table knowing the specific details of the mission parameters, including myself."

The director cleared his throat and shuffled again in his chair. Leonardo couldn't see his face, and he was looking at his reflection on the table again as he listened,

"I disagree with Leonardo," started the director, "and I say we go to a vote."

The director raised his hand, "all in favor of Leonardo *and* myself having access to the mission-critical information."

Leonardo looked around the table and saw some board members raising their hands.

"There you have it," said the director smugly. "Six votes, it's decided." He sat back in his chair, obviously pleased with himself.

"Wait a minute," Leonardo interrupted. "I count five votes, director. You can't vote for yourself. You and I should both be abstaining; that will make ten total votes, and you only have five."

Leonardo and the director's eyes locked. Leonardo could see anger in the director's eyes; *why did he want to know names and places? That will be boring, irrelevant information.* Leonardo examined the director's face closely; he had piercing black eyes and big bushy eyebrows to offset the long goatee, and his pock-marked and scarred facial features stood out to Leo as he examined the director closely in this revisited memory.

The sight of the director began to become blurry and then disappeared as Leonardo opened his eyes with a start.

Kevin nudged Leonardo again, and Leonardo

looked up to see the stewardess, who repeated her question, "Sir, I have a message for you; are you Stan?"

Leonardo sat bolt upright from the relaxed position in his seat. "Yes, I am!"

"Sorry, sir, but I was asked to give this to you when you first got seated in the plane, but I was so busy getting people on the plane and then doing my takeoff checklist. I guess I forgot," she paused, "I'm sorry." The stewardess sheepishly gave Leonardo a small folded-up desk note, turned, and walked back towards the plane's rear.

Leonardo clutched the folded paper to his chest and glanced at Kevin, who was looking at him with his eyebrows raised.

"Well? Stan, is that from our friend?" Kevin asked.

Leonardo leaned back as he started to unfold the paper. He began to sweat. *Was this from Bruce? Please let it be from Bruce.*

Leo read the printing on the napkin, and his heart sank as he started to sweat even more.

The note read,

Stan. Sorry, I will be late.

C.

Leonardo flipped the paper back and forth to see if there was any other information; he examined the airline crest on the top of the paper. Nothing more. He looked at the writing more closely; it was not Bruce's writing.

Leonardo looked up and saw Kevin looking down at the note. "What the heck does that mean?" Kevin finally said, still looking at the writing on the note paper.

"It means our friend didn't get on this flight like he was supposed to," stammered Leonardo. His mind was racing to think of what could have happened to Bruce that he didn't make the flight, but he was well enough to get this message to Leo on the flight.

"Who's C?" Kevin asked, still looking at Leonardo curiously.

Leonardo finally looked up to Kevin as he folded the paper and put it in his inner jacket pocket.

"C is my partner. Charlie." Leonardo said it as

casually as he could, looking at his watch as he put his head back on the headrest.

He had about four hours left on the flight and needed to reorganize his thoughts before landing in Honolulu. Things felt markedly worse without Bruce on the plane.

'C' stood for Champ.

B.E. Smith

Chapter 43: I Fought the Law

Bruce was walking down the side of the secluded highway in the dead of night. He hopped into the ditch and crouched as cars drove past him, oblivious. Although the long nighttime walk was unexpected, Bruce was thankful that he had escaped with almost all his possessions. The walk allowed his mind to process the day's events and devise a contingency plan to get to Hawaii as soon as possible.

As he walked, he hoped Leonardo got his message, and Leo and Kevin were safely on their way to Hawaii. Processing the day was a priority for Bruce's mental health, especially today; he had found over multiple missions that his psychiatrist Sandy's direction to replay the events back and then mentally sort them was the best way for Bruce to cope with unexpected ongoing events. He started back at that morning before all the wheels had fallen off:

Bruce watched from a distance as Leonardo Hoffman and Kevin Nault sped away in the Chevrolet Impala at 7:02 a.m.. Bruce observed Leonardo as he impersonated an FBI agent, and he was impressed that Leo had gotten Kevin into the car faster than expected. As the car zoomed down the street and around the corner, Bruce got to work, collecting the two nanocams he strategically placed; one was across the street from Kevin's house on a street lamp, and the other was a block away near a busier intersection.

Having collected the two nano cams, Bruce ran for about fifteen minutes north of Kevin's house as he came to two of the more populated parking lots in the area. The two lots were adjacent to the university campus, which was convenient to Bruce for several reasons. The primary reason was to steal a car, and the owner wouldn't realize their vehicle was gone until later that day. Many people were milling about, heading to work or school; Bruce watched from a park bench as the people went about their morning routines, oblivious.

He hadn't been sitting on the bench for five minutes when he saw a little red Volkswagen Rabbit pull into the lot and expertly pull into a spot about halfway down the first lane of the parking lot facing him. Bruce examined his watch as the nanocam he had attached to the rim of his ballcap scanned his field of view. As Bruce was tracking the little car with his eyes, AI began examining details about the vehicle that was important to him. His watch gave a detailed description:

Vehicle: 1980 Volkswagen Rabbit, vehicle condition: good, fuel: diesel, age of driver: 18 years old

As Bruce continued to watch the small car, AI kept analyzing,

Emissions: Diesel engine, sound, and chemical spectrometry show that the machine is working well.

The rabbit pulled into the parking spot, and the driver reached over to get her packsack and a thermos, exited the car, and briskly walked through the parking lot as she gazed at her watch. Bruce glanced at his watch as the young lady glanced at hers.

Prospect: good: manual door locks, manual window mechanism, manual engine startup.

Bruce got up from his temporary seat on the old, wood-slat bench, and his hand tool was ready as he calmly walked across the street to the parking lot. As he gazed left and right, AI gave a single vibration on his wrist, indicating that no nearby people were watching him.

As Bruce got between the two larger cars and to the driver's door of the rabbit, his muti-tool was up into the top rubber seal of the window, and as he got a finger into the space above the window, his tool then pried the bottom rubber of the window away. A quick lift and push down on the window disengaged the manual window mechanism, lowering it about six inches. One last quick movement as his skinny arm expertly reached in and pulled the manual locking mechanism up, and he had the driver's door open.

Bruce threw his backpack onto the front passenger seat as he sat down and pulled the panel from the bottom of the steering column. Bruce reached under the steering wheel and over to the right towards the ignition lock as the wires came into view. As he felt the subtle 'click,' he turned the key ignition module, and the car roared to life. Bruce then put the panel back under the steering column, put the manual transmission into reverse, and was letting out the clutch and backing out of his spot by the time one minute had elapsed. Bruce had a smile on his face as he shifted through the gears, and the little car started its unplanned journey.

Bruce was now on his way to LAX, with AI perched

to the peak of his ballcap; it was beautiful. Bruce left behind two nanocams; one was in Kevin Nault's now dormant office at the university, and the other nanocam in Bruce's house on Fremont Street was severely damaged by the structure fire. Bruce looked over at his backpack on the passenger seat, relieved that the theft was unnoticed. Chances were that the young woman whose car Bruce had stolen would only notice her car was gone likely after lunch, at which time Bruce would be in the State of California.

When he was leaving town, he marveled at the ad hoc plan he and Leonardo devised at the last minute, as their circumstances drastically changed with the murder of Miss Rosie and the destruction of Bruce's house. As Bruce got another notification by vibration from his watch, he asked AI to give notifications audibly as he drove down the highway; he wanted to be sure that the bait they had left for Shadow would have the best chance of success. A few minutes passed in silence before AI spoke, "Nanocam in Doctor Nault's office detecting movement and sound. Would you like to hear the audio?"

"Yes," replied Bruce.

The audio from Kevin's office began playing through as Bruce listened on,

"I've got this message from Kevin to photocopy two hundred of these lecture posters since he's going to be away unexpectedly until his lecture on Friday," the female voice said, "he said to come and check on his desk for a notebook he left here." The lady was rummaging through some papers on Kevin's desk, "Hmm, I don't see it."

"Did you check the staff mail room?" the second lady asked.

"No, I'll check there too, Karen. His faxed message says he already booked the big lecture hall at the Tucson campus. We need to leave about fifty flyers in each of the physics graduate students' mailboxes. They can pass them on to some physics students to put up around campus here."

"Okay," said Karen, "if you take care of photocopying the two hundred posters for this campus, I'll fax this copy of his lecture poster over to Valerie at the Tucson campus and ask them to put up posters at the Tucson campus."

Bruce smiled at the simplicity of Shadow's bait. It

was a quick scheme to get Shadow's attention, and whether Shadow became suspicious of the advertised lecture by Kevin Nault or not was irrelevant. Keeping Shadow guessing about Bruce and Kevin's location was paramount, and if Bruce had his way, he would be waiting at the gate for his plane to Hawaii by the time Shadow had come up with a course of action. Any way Bruce and Leo had figured, Shadow would likely not find out they were leaving the continent until later.

Your move, Shadow, Bruce thought as he sped down the minor highway toward the Interstate.

Several minutes later, Bruce merged the little car into the moderately heavy traffic as he started down Interstate Highway 10, and the little rabbit sputtered its diesel exhaust fumes. AI chimed when he got onto the I-10 and gave an estimate for the drive time to LAX.

Bruce would have time to stop and get gas once before he abandoned the little car outside the airport. He thought about Leo and Kevin and hoped their travels were going well; they were only about thirty minutes ahead of Bruce.

Interstate 10, which traversed the distance from Arizona to California, was a well-maintained highway with two lanes for traffic in each direction. The highway wouldn't see a lane increase until 2007 as the population density grew, and multiple suburbs sprung up across the desert. Bruce started to hum as he drove amongst the significantly larger vehicles going West. He was following the average traffic speed for several miles when he noticed a police cruiser coming up from the rear of his traffic group.

Bruce remained calm as he glanced over and over in his rear-view mirror; the police cruiser was getting closer and closer. As vehicles in the fast lane began to move to the right to allow the police car to pass, it just stayed in the slow lane two cars behind Bruce. He began to formulate a plan in the unlikely event he would get stopped by the police; Bruce had his fake Arizona driver's license, and he hoped that the vehicle documents were in the glove box as he was looking around the tidy car and didn't see any documentation laying around.

Bruce squirmed in his seat as he surreptitiously watched the police car. As the group of vehicles continued West, an interstate exit sign passed, and Bruce saw the car between himself and the police cruiser pull into the right exit lane.

"Come on, please exit," Bruce muttered under his breath as the police cruiser seemed to be cemented into the traffic lane now directly behind Bruce. The exit lane ended as Bruce continued west; now, he felt his heart rate going up, and he could see the police officer talking on the CB radio. Without moving his torso or head, Bruce reached over to the passenger seat with his long arm, undid the zipper of his school bag, and grabbed the small black vial containing the nanotech.

With his arm returning to his side, Bruce scratched his ear with his right hand before putting his arm down and sliding the black vial into the band of his underwear. Once his arm was down in a low position with his hand still on the top of the black vial, he could verbally command AI to attach one of the nano weapons to his index finger.

Bruce weighed his options, and it didn't take long for him to decide to use the Neurosynaptic Dampening Injector only if all other non-violent techniques had failed. Although he had several NDIs on multiple nanobots, he did not want to injure a police officer; it was wrong, and he did not want that kind of attention. As the next exit approached, Bruce got ready to exit and hopefully lose the police cruiser. He signaled right, moved into the exit lane, and watched as the police cruiser stayed in its lane going West. He slowed in the little red car as the off-ramp curve circled to the right, and he started to get out of view of the police cruiser.

Just as Bruce was sighing a big breath of relief, he saw the police cruiser, with its emergency lights on, coming up behind him fast. Bruce jerked the steering wheel to the right of the roadway just as the police cruiser sped past him and off the ramp onto the busy street that lay ahead. Bruce was almost stopped on the off-ramp, over to the right of the lane on the last part of the bend before the straightaway into town. A few cars passed him before he crept back into the live lane so he did not stick out.

He instructed AI for the nanobot on his finger to disengage and return to the storage container he now held open with his right hand. As the nanobot returned to its home, the little red car sputtered into life as Bruce accelerated around the last part of the curve and downhill on the straight roadway. He could see the police car stopped partway down the road with its single rotary emergency light still activated on the car's roof.

As Bruce drove down the street and past the empty police cruiser, which was parked awkwardly at the side of the busy boulevard, he sighed a breath of relief again as the traffic light ahead of him turned yellow and red. He pushed the clutch in and stopped at the red light, asking AI for directions to get back onto Interstate 90.

AI began verbally giving directions, and as Bruce glanced at his watch, he didn't see the man run up to the driver's door of the red rabbit and open it, pulling Bruce partway out of the car in the process, since his hand had been resting on the inside door handle.

The man began yelling at Bruce and immediately brandished a revolver, "Back in the car carrot top!"

Bruce had been caught entirely off guard and was off balance. He had been watching the police cruiser intently and then focused on getting back onto the Interstate.

As Bruce reacted to the assault, he widened his stance and squared his body to face the man. Instinctively,

Bruce thrust out his left hand and pushed the right hand of the assailant, which was holding the gun, and as he did, the revolver fired off to Bruce's right side just as Bruce's right hand whizzed in straight on to deliver a devastating palm strike on the man's lower jaw. As Bruce's left hand came back to provide a second blow to the face of the man, he paused as the man's eyes had fluttered closed, and he reeled backward as his nose gushed blood and he fell in a heap with a loud thud onto the boulevard.

Bruce instinctively put his foot on the revolver, which had fallen to the ground after it fired, and the man had been rendered unconscious. Bruce dragged the gun from his right side to sit under his foot.

Bruce could hear gasps all around him as the wholly unexpected events unfolded around him, and as the man lay in a contorted heap on the edge of the roadway, Bruce's attention left the man. As he looked up, he saw two policemen standing on the curb, their firearms drawn and resting at their sides. One of the police officers reached down and flipped the unconscious man onto his stomach, handcuffed him, and left him lying in a pool of blood. The second police officer approached Bruce and asked, "Are you okay, sir?"

"Yes, I'm fine, sir," Bruce exclaimed, astonished at how the events had transpired to have him face-to-face with the police officer he had been trying to avoid.

"I've never seen anyone react that fast. Are you police or military?" the officer asked, somewhat confused at Bruce's handiwork.

"Uh, no sir, I'm not, but I have studied martial arts since I was a kid," Bruce said as he lifted his large foot off the revolver still under his foot. "Do you want the gun?" Bruce asked.

"Yes, son, thanks," the first police officer grunted as he came up and reached down to grab the gun off the street. Bruce watched in amazement as the officer reached down and grabbed the gun, with no glove on. He expertly unloaded it, placing the remaining round and the spent casing in his side cargo pocket. The officer stood up and looked past the little red car still idling there, looking in the direction that the man fired the revolver.

"It should have hit down there," Bruce said as he pointed with his long arm down towards the hood of his car,

"I think I see where the round hit," Bruce followed up.

The burly police officer ambled over to where Bruce had been pointing and leaned down to see the bullet hole in the hood of Bruce's car. The car sputtered and made funny noises as the two men looked at it, and then it gave a last sputter and stopped.

"Where are you from?" the police officer asked quizzically. He scratched his head at the situation he found himself in, with this tall, lanky young man who was hyper-aware of his surroundings and unusually calm about the ordeal.

"Uh, I live in Willcox, sir," Bruce replied. You *don't want to know where I'm from*, Bruce thought.

The police officer removed his patrol hat and scratched his balding scalp. "What's your name, son?"

"George Cooper, sir," Bruce lied. He took a breath to talk, but the policeman spoke first.

"I need you to come with me to give a statement about what happened. I'll give you a ride since your car's quit."

That's just great, Bruce thought as he looked at the little trail of smoke coming from under the hood of his stolen car. He watched as the other police officer lifted the groaning man off the pavement by the chain between the handcuffs and dragged him to one of the police cruisers nearby.

"Over here, Mr. Cooper," the other policeman shouted at Bruce as he got into a second cruiser that must have arrived after Bruce had driven by.

Bruce grabbed his backpack from the passenger seat of the bullet-wounded car and looked around before he used the bottom of his t-shirt to wipe the steering wheel, the ignition crank, and the door handle if they tried to find fingerprints in the car. He turned and slung his backpack over his shoulder as he walked towards the police cruiser; this was not what he was expecting, although being offered a ride as a witness in the police cruiser was far better than being put in the back of the cruiser for stealing a car. Bruce took some deep breaths as he got into the front seat of the police cruiser, which the officer had opened for him.

"Shouldn't take too long, Mr. Cooper," the burly policeman grumbled as he took off his peak cap and hung it over the double-barrel shotgun strapped between the seats, "we just need to get your statement at the station; you want us to call your parents?" he continued.

"No, no, sir, my parents would just worry, and besides, I'm fine. No need to worry them for nothing," Bruce interjected.

"Okay, son, you're old enough to decide, and I would think that a young man with your defensive talents doesn't need me to call his mommy," the policeman chuckled before continuing his admiration of the knockout punch, "You were so fast, I didn't even get my gun drawn before you cold-cocked that guy, the blood was gushin' out his nose even before he went down like a sack of potatoes! Couldn't have done it better myself! You interested in applying for a police constable job, son?"

"Uh, not right now, sir, I'm still in school," Bruce replied as he looked at the sergeant stripes on the muscular man's shoulders and then down at the officer's name tag, which read, 'James.'

"Ah, yes, good idea. Get your schoolin' done first," Sergeant James replied.

"Yes, sir, and if I want to get into policing after that, I'll certainly contact you, Sergeant James," Bruce lied as he continued to plan what he would do to get out of the police station unscathed.

Sergeant James was on the CB radio after some chatter about Bruce's car blocking the intersection, "Yeah, Sergeant James here, that car can be towed to the tow yard, and my witness can decide what he's doing with it later." He glanced over at Bruce as he spoke, "Sounds like your car is toast, George, but I'll give you a lift wherever you want to go when we finish your statement. If it's not too far, mind you. You can come to an arrangement with the tow company later."

"Yes, sir," Bruce replied, "I think I'll leave the car there until I can get my parents to help me go get it. I would appreciate it if you could drop me off at the Greyhound bus terminal when we are finished."

"Where you going?" Sergeant James asked.

"I was headed to have coffee with my friend when this happened, but I think I'll get the bus back to Roadforks, where my folks live. I'd like to see my parents after what happened this morning." Bruce fibbed.

"Sounds like a plan," Sergeant James grumbled as they pulled up to the police station.

Looks like I'm going into the wolves' den before I can get out of here, Bruce thought.

Chapter 44: Leo Steps Up to the Plate

Shortly after 11 p.m., the 747 touched down at Honolulu International Airport. A short taxi to the arrival gate took about five minutes as Leo and Kevin shuffled in their seats and grabbed their briefcases.

As the cabin door opened, the fresh, moist sea air rushed into the cabin, temporarily pushing the smoke-ridden air from the front of the plane towards the back. Then, as the air cleared, Leo breathed a sigh of relief as he took in the sea-level oxygen-laden air in a few long, deep breaths.

This has to go right; stick to the plan, Leo, Leonardo thought, as he shuffled towards the front of the plane in the slow, funneled movement of people. He clutched his briefcase close to his chest as he walked closely on Kevin's heels; he wasn't taking any chances with any aspect of the plan, and dropping or losing the vital contents of the briefcase wasn't going to happen. As they got to the front of the plane, Kevin smiled at the stewardess and nodded as he turned to get onto the walkway into the airport terminal. Leonardo did the same, giving the stewardess a barely audible "Thanks" as he walked by.

"Have a good day, sir," the stewardess said to Leonardo as he walked past her. The same lady had given Stan the napkin message, and Leonardo touched his suit jacket pocket with his free hand to mentally check that the message was still there.

Both men followed the stream of people up the walkway and into the Honolulu airport arrival area, and they made a beeline to the signage labeled 'Exit.' The men continued walking in silence until they exited the airport and into the fresh sea-level air; Leonardo inhaled deeply again, favoring the fresh sea air.

Leonardo got right back into character,

"So, Kevin, what's the address we're going to again? I'll get us a cab."

Kevin's reply was quick, "What do you mean? Don't you know? We should ask to go to the telescope; it's the only one on the island, and the researcher's quarters must be close by." Kevin was still crinkling some papers in his hand as he spoke. "I thought we were going to an FBI safe house?" Kevin looked at Leonardo with his head tilted.

"It will be a safe house for us, but I can't elaborate more than that, Kevin," Leonardo quipped. "And you can't say those words in public. You could draw unwanted attention to us; this is more serious than you think, Kevin." Leonardo continued in a more disciplinary tone.

"Alright, Stan, you're the boss, I'm just asking," replied Kevin awkwardly.

An orange cab approached the two men, and Leonardo leaned in to speak to the driver through the open front passenger window.

"Hi there, we want to go to the big telescope. Can you take us there?" Leonardo wasn't sure what the local people called the telescope.

"Ah, yes, sir, the big telescope up on Mauna kea," the man sang in his thick Hawaiian accent. He was dressed in a stereotypical flowered yellow shirt and had a ball cap sitting atop his curly salt-and-pepper hair. The man shifted the taxi into park, got out, and opened the trunk of the car for baggage, but as Leonardo directed Kevin into the rear seat of the cab, he went back and spoke to the cabbie,

"No big bags, sir, just our carry-on luggage."

"Ah ha, okay," the man laughed, closed the trunk, and returned to the driver's seat.

Leonardo got into the front passenger seat, and as he closed the door and visually checked that he and Kevin both had their briefcases, he turned and smiled at the cabbie, "We are here just for a quick visit, so we don't have many bags."

"Yes sir, I hope you enjoy Hawaii; so much to do, and the weather is always good," he laughed a belly laugh as they drove away from the airport. Leonardo examined the cab driver and marveled at the stress-free man, smiling and looking around as he drove and started to hum to himself.

I don't remember a time that I felt like that, Leonardo thought. The decades of academia and the high-stress environment of his quantum research came to his mind, and he couldn't think of a time in his adult life when he felt completely at ease or stress-free. Leo loved his work and was extremely passionate and driven, but that took a toll on the body and the mind. He moved his head back to look forward as he realized he was staring at the cabbie as he thought.

"Is there a place to get food near the telescope?" Leonardo asked the driver. He looked at the taxi permit on the dash with the driver's picture and the name Keakaokalani.

"Yes, they have a little tuck shop near the telescope. We get calls to bring fresh food and meats up there occasionally; gotta keep those researchers happy, no?" he laughed.

"Excellent," Leonardo smiled at the man as they started to ascend the landscape. Both Leonardo and Kevin looked out the windows of the cab at the lights of the city getting farther and farther away.

"Actually, I think they closed the tuck shop for repairs. Hopefully, it'll be open again soon?" the cabbie smiled. "You do nighttime or daytime research at the telescope?" the cabbie smiled at Leonardo.

"Mostly nighttime?" Leonardo smiled back at

Kevin, "I'm guessing?" Kevin nodded at Leonardo and smiled a suspicious smile.

"How do you pronounce your name?" Leonardo quickly changed the subject.

"Kea-kao-Kalani," the man smiled. "I know it is a mouth full for white people; call me Kea-Kao. It means the Heavenly Shadow."

Leonardo flinched at the mention of Shadow but smirked at the cabbie, "that's a great name, Kea-Kao," he continued, getting the pronunciation pretty well on his first try. Having lived and worked most of his life in an academic setting, Leonardo Hoffman was used to multicultural names, was proficient in multiple languages, and could get by speaking several others.

The orange cab wound its way around and around as the vegetation changed from lush to sparse to barren on the Mauna Kea mountainside. The two travelers sat in silence, each one pondering his reality related to the unique circumstances that brought them together under the clear and starry Hawaii night.

Chapter 45: Getting Out of the Wolves' Den

Bruce sat in the steel-armed chair in the interview room, leaning over with his head, and his chest hung over his legs as he stared at his shoes. The interview had gone as seamlessly as Bruce had hoped; he had stuck to all the facts he said to Sergeant James on the way to the police station, so there was no reason for the young officer who interviewed Bruce to disbelieve anything he said in his statement. The constable, who had finished writing down Bruce's statement several minutes earlier, thanked Bruce for his cooperation and left the interview room to check with Sergeant James to see if there was anything else they needed from Bruce before they let him go.

In Bruce's mind, he was six steps ahead of where he was, sitting in the wolves' den, waiting to be let out to continue his journey to California and get a flight to Hawaii to join Leonardo and Kevin. Bruce made the brief call to the American Airlines desk from the payphone in the lobby of the police station near Leo's departure gate before his interview. He stood close to the phone and spoke quietly to ensure any lobby camera didn't capture his plan. With his incapacitated stolen car and being stuck in a police station, he left a message for Leonardo. He wondered if Leo would get the memo.

Bruce was rhythmically breathing as he stared at his shoes, trying to calm his body as he planned to escape his situation. The past few days had been out of the ordinary to the extreme: the telescope explosion, the murder of Miss Rosie, and the burning of his house on Fremont Street, not to mention the extraction of doctor Kevin Nault and Bruce's chance altercation with a gun-wielding man who had made the grave mistake of trying to hijack Bruce's little car. The one he had stolen.

It was a thick stew of circumstances that would make anybody's head spin, but Bruce Hayden had taken it in stride thus far, improvising and adapting at every change in events.

He was musing how the wait in the police station interview room had been the only time in the last several days that he could sit and think without anywhere to go, although he didn't have a choice.

The door to the interview room boomed open, and Bruce sat bolt upright and smiled his goofy teenage smile as Constable McTavish came in, studying the papers in his hand as he turned and closed the door behind him and sat in his chair, putting the folder of documents on the table. He looked up and smiled back at Bruce.

"So, that's all I need right now, George, and if I need anything else, I'll just contact you at your parent's house."

"Yes, sir, I'm not going anywhere. If you need anything else, I'm happy to help," Bruce lied.

"Well, okay then, I'll get my patrol stuff together, and I'll pick you up at the front of the station and drive you to the bus stop so you can get back to your parent's house." Constable McTavish stood up and shook Bruce's hand, "Nice to meet you, young man, and please keep us in mind when you start looking for a career. Sergeant James is just smitten with your handy work; that guy you knocked out will be eating through a straw for weeks," he chuckled, "but don't tell the sergeant about the smitten part," he smiled.

"No sir, I won't say a word. I was doing my part to

help out." It was the first truthful statement Bruce had made in a while. Well, mostly the truth; the truth was more like, 'I won't say a word, so you let me out of here, and I was just protecting myself from that psycho with the gun because I have a date to discover an asteroid and save the planet.'

Bruce walked out into the hallway of the stuffy police station and was shown to the front lobby of the station. As he walked into the lobby, the constable turned on the lights; it was almost dark outside.

"You can wait here or outside; I'll be a few minutes and meet you out front."

"Yes, sir," Bruce said as the constable disappeared back into the station. Bruce headed immediately outside to get some deep breaths of fresh air.

'Freedom - almost,' Bruce thought as he paced back and forth in the parking lot with his backpack over his shoulder and his ballcap back on his head. He sighed a few breaths of relief as he focused on what he had to do next, and the police cruiser came around from the rear of the police station and stopped near Bruce. The evening twilight was relieving the waning daylight as Bruce got into the front of the police cruiser, and they began traveling east, the opposite direction that Bruce needed to go.

No words were exchanged during the twenty-minute trip to the Greyhound bus stop outside town. The police cruiser kicked up the gravel at the side of the highway as it slowed to a halt, and the red-haired teen got out and closed the door, slinging his backpack, saying, "Thanks!" as he closed the door.

The police cruiser drove down the highway a short distance, and then did a U-turn and came back and drove past Bruce, its headlights quickening down the road as the vehicle accelerated out of Bruce's life. That is what Bruce had hoped at the time.

As soon as the police cruiser was out of sight, Bruce activated AI, which was still perched on the peak of his ball cap.

"AI, scan my surroundings as I run, and give me updates by audio. Same as before, if another person is nearby, mute audio."

"Confirmed," AI responded, "You are eight kilometers from the bus station where you can travel West to California. At the next intersection, turn south." AI finished.

"Yeah, I don't want to stay on this roadway in case they become suspicious," Bruce mumbled as he started into his rhythmic running gait and coasted down the desolate highway.

AI interjected, "As a result of monitoring all of the conversations since this morning, I have extrapolated that the sergeant suspects something. I have calculated the chances to be 62% that he arranged to have you dropped off at this secluded bus stop because he is suspicious. It could be interpreted as forcing you to stick to the information you gave them about your destination. The police officer could have dropped you off at the closest terminal building, about ten minutes from the station."

"Yeah, that's calling my bluff," Bruce laughed at AI's wordy explanation of the sergeant's possible tactic.

"I was wondering that myself," Bruce said as he ran down the dark highway.

A few minutes later, Bruce felt much better; the endorphins from running started to ease the unusually high stress he had been feeling for days.

"What is Leonardo's location?" Bruce asked AI as he rounded the first corner to head south and follow AI's suggested route.

"Doctor Leonardo Hoffman's location is most likely over the Pacific Ocean en route to the Island of Hawaii," AI chimed, "his AI unit burst a signal in the five Gigahertz band as his plane was leaving the area at three thousand meters. I received it forty-eight minutes ago while you were waiting in the interview room," AI finished.

"Haha, Leo, you genius! Yes! Great job, Leo!" Bruce yelled as he thrust his way down the middle of the highway. He laughed at Leonardo's smarts while in the field; Leo never admitted it, but he could adapt and conform to many pressurized situations. Now, he had proven himself multiple times in the field that day.

Making Bruce's nanocam unit a temporary stand-alone AI by replicating the complete AI unit was all Leo's idea; It was the last thing that Leonardo insisted on doing before he and Bruce separated that morning. The conversation played back in Bruce's mind as he ran down the dark highway.

"Leo, I know why you want to split the AI into two units; we are just running out of time. Kevin leaves for work in less than an hour." Bruce was frustrated with Leo, who was moving much slower than Bruce was used to when the stakes were high.

"Bruce, just get yourself ready and make sure my fake ID is good for the flight to Hawaii. It'll take me 20 minutes at most to copy this unit and weave a temporary AI into your nanocam's memory. Nothing left to chance, right?" Leonardo had already snatched the ballcap off Bruce's head and placed it in the briefcase next to his AI unit. Bruce returned to Leo's wallet, which was on the bed, and started double-checking all the information and credentials Leo would use that morning.

Leonardo rubbed his hands together as he focused on the task at hand, "AI, replicate user interface, along with global position tracker, interactive scanner in visual, infrared, and alpha, beta, and gamma bands of radiation. Upload maps to LAX, all vehicle specs and spectra information, and create an audio interface replica to be activated by Bruce Hayden." Leo tapped the desk beside the ballcap, "Replicate to the nanocam weapons unit mounted to this ball cap."

"Copying," AI responded as it silently copied the parts of itself to the nanobot's structure and memory on Bruce's ballcap. AI continued, "Estimated time to weave stand-alone AI unit, twenty-five minutes."

Leonardo glanced over at Bruce, expecting a protest at the time it would take. Still, Bruce was ensconced in taking everything out of Leo's wallet and briefcase and repacking it to ensure the LAX experience was successful.

AI instructed the nanobot to build a structure and electronic components using spare nano exoskeleton pieces that were part of the existing nanobot. It could modify itself by repositioning its components to satisfy Doctor Hoffman's request. Leo thought momentarily before speaking to AI again, "AI, what will the battery life of the new AI stand-alone unit be, assuming normal use and constant threat scanning?"

"Battery life approximately six days, given normal usage," AI responded.

Leonardo quieted his voice for his next question for

AI, and it silently copied its user interface to the memory banks of the nanobot attached to Bruce's ballcap. The nanobot rearranged its structure to match. "AI, how long would it take to add the capability to message between AI units over a long distance?"

"Five minutes to encode, however, messages broadcast at five Gigahertz will significantly decrease the battery life of the temporary AI unit, especially over distances larger than 4.83 kilometers, which is the line of sight. Farther than 4.83 kilometers will require an object to reflect the signal...."

"Right, stop," Leonardo cut off the AI, shaking his head at the stupidity of thinking that there was satellite communication, "If we aren't in the line of sight and can't get up to a higher elevation, it won't work."

"Correct," AI chimed, "however, when the moon is properly positioned, I can reflect a signal off of the lunar laser ranging retroreflector array installed on July 21, 1969. That would translate to a signal broadcast distance of approximately 4,000 km, depending on the moon's location in its orbit around the Earth."

"Leo," Bruce interrupted, "you have ten minutes until you have to leave."

"Yes, yes," Leo responded, getting up to check his suit in the mirror as AI continued to modify Bruce's nanocam unit.

"Time to completion: twelve minutes," AI said as it followed the conversation in the motel room.

"Leo," Bruce warned.

"AI, re-optimize tasks, and never mind the independent communication between units. Just leave the capability to receive incoming five Gigahertz signals," Leonardo said as he nervously straightened his tie. "That should keep the battery usage low if we have to get you a message."

"Minimal negative effect on battery consumption. Time to completion: eleven minutes," AI responded.

"Leo, if you're late to get Kevin, you'll chase him in his car. He leaves at almost the exact time every day. 7:00," Bruce growled.

"Yes, yes," Leonardo huffed. I'll get there on time; I want your AI unit to be functional for my peace of mind."

"Okay, Leo, I'll finish repacking your briefcase and wallet and get your car ready to go; just make sure you're zipping out of here the second AI is finished."

Bruce and Leo spent the next ten minutes in silence, except for AI announcing every minute as it counted down to zero. Bruce handed Leo his wallet with fake FBI identification, and Leo practiced flipping it open in front of the mirror.

"Detective Jones, FBI.... ahem...Detective Jones, FBI," Leo repeated over and over with the same flip of his wallet to show the fake identification, "you think I should say it slowly or at a normal speed?" asked Leo as Bruce packed up his briefcase.

"You sound like you're trying too hard, Leo. If you believe what you're saying, it'll come out better. Try not to think of who you're impersonating but who you need to be for this mission. You need to be Detective Stan Jones, Leo; you're saving Kevin's life by doing this.

You gotta own it, or Kevin will be suspicious. He's a sharp, sharp man. And these FBI guys say that so many times that it loses its luster. Try saying it like you're exhausted and just walked out of a room full of dead bodies."

"That's oddly specific, Bruce, but okay," Leo cleared his throat, re-set his wallet, and put it in his inside suit jacket pocket. "Detective Jones, FBI." Leonardo flipped the wallet open and announced himself in the most tired and stressed way possible.

"Much better, Leo," Bruce responded excitedly, "do you feel the difference?"

"I think so," Leo responded. He still looked himself up and down in the scratched, skinny mirror.

"Remind me that when we are out of dodge, I'm buying you a nice dinner, young man. I had a pretty good idea of what you did on these missions, but being in it has an entirely different flavor.

"I know, right?" Bruce announced with his arms up as he threw the last neatly folded shirt into Leo's briefcase and placed the package of papers on top, "You get used to it, though, Leo, you'll see. Just don't let your guard down for one second. Keep saying to yourself, Detective Jones, FBI, or Stan Jones. For the foreseeable future, you're not Leonardo Hoffman; you're Stan Jones. You can't slip up even once."

"Yes, yes," was Leonardo's predictable response.

Leonardo was pacing at the door with his briefcase in hand when he heard AI counting down from its temporary home inside his briefcase. As AI said, "Completed," Bruce and Leonardo ran out the motel's door. Bruce dove into the rear seat and watched as Leonardo threw his briefcase on the front passenger seat, put the car into drive, and sped out of the motel parking lot as AI called out directions from the passenger seat.

AI snapped Bruce out of his daydream as it called out the next intersection ahead of Bruce, who was nicely into his running rhythm, "next intersection, turn west," it announced.

Bruce was smiling as he thought of Leonardo sitting next to Kevin on the plane for nearly six hours; it felt like a shame that nobody else would ever capture the significance of the two scientists being together in person.

Bruce would never admit it to Leonardo, but they were the two most important and influential people in Bruce's young life and, arguably, in modern history.

Bruce had been running at his steady gait for twenty-five minutes when AI announced, "Vehicle approaching from the rear."

Bruce glanced behind him as he could see the distant headlights coming over the crest of the last small hill he had run past. It was still about a kilometer away, and Bruce would never have been seen in the shadowy surroundings if he had gotten off the roadway in time. Bruce slowed to a walk and got off the highway into the shallow ditch; he removed his backpack and lay down on his stomach with his bag on the ground before him, resting his hands on the pack. The ditch was dry, and Bruce briefly examined the intricate cracks in the soil that had not been wet for some time.

Bruce waited on the ground as he heard the car

traveling up the roadway; it seemed to be going significantly slower than he would expect on a back route.

AI reported the details from the approaching vehicle as Bruce peeked his head up from his prone position and looked back to where the vehicle was coming from. "Vehicle is three hundred eighty-nine meters, closing at fifty-six kilometers per hour. The vehicle has a secondary light source consistent with a side-mounted police searchlight."

Bruce's heart sank as he got back on his stomach and put his ballcap on the ground facing the roadway. "Will I be visible if he's using a searchlight?" Bruce asked AI as he contemplated running for it into the desert.

"Negative. Given the speed of the car and the depth of this ditch, it is unlikely you'll be seen," was AI's quick response.

"Unlikely is not comforting right now," Bruce thought as he lay still in the trough of the sandy ditch.

The car continued its slow approach, and Bruce told AI to be silent. He waited as he heard the car get closer and closer. Bruce's breathing slowed as he strained to hear anything besides the car tires. The area to the right of Bruce up on the other side of the ditch lit up, and Bruce watched the light pass by and then move towards the roadway again. He could hear voices from the police radio as the cruiser came even closer to Bruce's hiding spot.

"Lima 2-0-2, do you copy Lima 0-0-2?" The windows were down on that warm Arizona night, and although Bruce could not make out everything being said on the radio as the car approached, he did hear the unmistakable booming voice of Sergeant James on the radio.

"Yeah, this is Sergeant James. I'm at the impound yard with that little red car. It is the stolen car we talked about."

Bruce's eyes shut briefly as his fear was confirmed: George Cooper would have to disappear. Fast.

Sergeant James continued, "I want three of you out there to block off the two roads back into town. The kid couldn't have gone far." The rest became mumbled, but Bruce held up his ballcap off the ground, AI pointed at the police cruiser as it continued past him, the darkness trailing the car like the wake of a large ship.

When the cruiser was well away from him, Bruce got up and sat cross-legged at the top of the ditch as he watched the police cruiser continue away. "AI, give me some optimized courses of action with parameters observed."

"Calculating." AI took no more than a second to respond again, "Option one – travel across open land due West from your position back into town and steal another car to escape the zone. Option two…"

"That works for me," Bruce said as he cut off AI, slung his backpack, and started at a trot across the dark desert landscape. "AI, call out any obstacles or divots in the soil larger than five centimeters."

"Affirmative." AI silently began scanning the landscape ahead of Bruce in multiple wavelengths as he forged a path fast enough to impress an Olympian.

The running style known as 'boot running,' perfected in the military for running over uneven and rugged terrain, came second nature to Bruce and adapted nicely to the contoured countryside.

The technique of running while the feet are shuffled along proved advantageous as Bruce made excellent progress across the barren ground; his feet never came high off the ground, which kept the body balanced and reduced injuries if any unexpected uneven terrain was encountered. AI kept up nicely, adapting to Bruce's changes in direction to navigate around changes in the landscape, and didn't announce obstacles that it knew Bruce already had seen.

After several minutes of running flat out, AI interrupted Bruce's rhythm, "Distance to the nearest neighborhood, four point six kilometers due West."

"Copy," Bruce replied; as he had reverted to his military-type training, the lingo came to him automatically. Bruce spoke again, "AI, as we get closer to that neighborhood, scan for a path into the neighborhood without using the roadway and scan for a vehicle that would be quick for me to steal. And I want to know if any police cruisers come within fifteen hundred meters of me."

"Copy," AI replied.

Bruce thought of how much Sergeant James wanted him captured; he had lied to him and completely fooled him into his identity. The old school sergeant, like Sergeant James, did not take being made a fool lightly at all. Bruce had an idea.

"AI, when we get closer to that neighborhood, dispatch one of the weaponized nanobots as far South as would be realistic for me to have traveled on foot and target a decent-sized vehicle parked on an unoccupied roadway.

I want the nanobot to turn that vehicle into a fire and light show. Time for the explosion and fire to coincide with me getting to the opposite side of the neighborhood to the north."

"Copy," AI replied, "time to fire a weaponized nanobot is approximately two minutes, depending on your speed."

"And AI, I want that nanobot coming back after it discharged its payload to start that fire."

"Copy," AI answered.

Bruce picked up his speed as he could see the nearby neighborhood come more and more into view. It did not seem like two minutes to Bruce when AI announced the launch of the nanobot.

"Launching" AI announced, "flight time to target approximately ninety seconds."

Bruce smiled at the simplicity of his distraction technique; he was confident that the police looking for him would flood the area of the car fire as Bruce slipped through the opposite side of the neighborhood.

Chapter 46: 1980's Policing

Like many decades before and since, the policing organizations felt they had a good handle on how to do their jobs in the 1980s. Comparatively with their future counterparts, policing in the 1980s was almost all stick and no finesse; it lacked some significant technologies such as DNA, video recording, drones, and nanobots.

Finding suspects and charging them with offenses was a grey area based on biased interviewing techniques and was rife with racism and homophobia. Or, in Bruce's case, having the suspect fall at your feet unconscious, like what had happened to the guy who had made the unfortunate mistake of carjacking Bruce's little, red, stolen car earlier that day.

Sergeant James stood on the roadway, smoking a large cigar while studying the flame-filled Buick. Minutes earlier, the flames rushed several meters above the large car as the fabrics, rubber tires, and large gasoline tank offered their potential energy for the light show.

"We're going to get nothing from this car, you know," Sergeant James mumbled to the significantly smaller constable beside him.

"We have five units flooding the area, Sarge. We'll find him," the constable replied, "and fire should be here any minute," he continued.

"Yeah, they can't water down this roadway until we get the dog here to get a proper scent," Sergeant James took a puff of his cigar, pulled it out of his mouth, and looked at the slowly decreasing saliva-soaked nub before speaking again to the constable still standing beside him. "You get going, so we'll have six plus me looking for this kid." Sergeant James took another long drag on the overused cigar. "I can't wait to get my hands around that kid's throat. He will regret this." Sergeant James' gaze left the burning car and landed angrily on the constable.

"Get!" Sergeant James yelled as the constable jumped and ran for his cruiser.

Sergeant James pulled his radio off his belt again and grumbled, "Sergeant James here. What's the ETA for the dog?"

The sound of another young male voice flooded the street via the radio still in Sergeant James' hand. "Yeah, Sarge, this is Fredricks. I'm about five minutes away; I'll come right to you and start the dog off at the car to get a scent," another radio officer announced.

Sergeant James did not respond but stood across from the burning car as he fumed about the day's events that saw this young man amazingly foil an attempted hijacker and then shortly after become a wanted man. He flicked his cigar between his large, sausage-shaped fingers; the sergeant would have to answer his boss about this mess, and he did not like being duped. He needed to have a suspect in custody before daylight, or he would wear this.

Minutes later, a police car whizzed around the corner, its lights flashing, and as the cruiser screeched to a halt about twenty meters from the burning vehicle, a young man all in green jumped out of the driver's seat and straightened his green ballcap as he opened the rear seat area of the cruiser. A primarily black Belgian Malinois jumped out of the back seat and sat, waiting for the officer to secure it to the extended search lead.

Once secured, the handler brought the dog reasonably close to the car and gave the command, "Find it!" as he circled the car in larger and larger concentric circles as the dog weaved back and forth with its nose on the ground as it covered the ever-expanding area around the burning vehicle. The dog was coming closer and closer to Sergeant James as the handler nervously looked on as he followed the determined dog.

"Head's up, Sarge!" the canine handler announced as the dog weaved its way closer and closer to the sergeant. Sergeant James stood there stalk still with the small remnants of his cigar in the corner of his mouth.

"I ain't movin'," grumbled Sergeant James out of the corner of his mouth that wasn't holding the cigar. The dog came closer and closer to Sergeant James, and just as the handler began to pull back and reign in his dog so as not to invade the stubborn man's space, the dog shot off to the right, away from the car with its nose on the ground. The pattern of weaving its nose back and forth, searching, became a straighter line as the dog pulled the handler away from the sergeant and up the middle of the street.

"He's got something!" the handler announced as he let the slack go on the long lead and began his well-practiced trot to keep up with his partner.

The dog pulled the handler up the street about fifty meters before stopping to circle again in his message that he had lost the scent. The handler stopped and let the dog circle as he encouraged the dog, "Good boy! Find it!" The dog circled wider and wider until it darted off in a direction across a residential front lawn, through a back yard, and onto the next residential street.

"Headed North from the fire scene through the yard at 386 Campion," the handler called out the house number on his radio as he followed the dog through the yard and onto the next street. The handler could hear the police cruiser engines winding up as they sped to attempt to contain the now-moving containment area. As the handler got out into the open street, the dog began barking frantically, and the handler saw a tall, lanky, darkly dressed male running up the roadway away from the dog.

"Police! Stop, or I'll let the dog go!" the handler yelled as the dog's barking increased in intensity.

The handler watched the man run across the street and into another yard, still heading North, as he started to run behind the agitated dog, still clipped to the long search lead. As the handler got to the front of the house where he had seen the male, he heard the unmistakable clinking of a person scaling a chain-link fence. He could see the fence in the house's backyard moving back and forth as the suspect climbed the fence out of sight.

As he caught up to his dog, the handler unclipped it and yelled, "Get him!" The dog became a black streak running across the yard like he had been shot out of a cannon. The handler had barely gotten in sight of the whole chain-link fence as he watched the dog jump the fence, having slightly broken its speedy stride. As it landed, it took two more leaps before there was a blood-curdling scream.

"Ahhhhhhh! Stop it! I give up!" yelled the young man as the dog slammed him into the grass on the other side of the fence and was still clamped down on his arm as the man yelled in agony.

As the canine officer jumped the fence and got within range of the suspect, he yelled at the dog, "Off!" as the dog let go of the man's arm and began barking at the man as he stood over the man's face.

"Ahhhhhh! Call off your dog!" the man yelled as he tried to shield his face and neck from the wildly barking dog.

The canine officer had just updated the other officers via the radio before he went in to handcuff the male, with the dog still hysterically barking nearby.

Two other officers ran into the backyard area from the next street over and stopped short of where the man was lying in the grass; the man's arm was bleeding badly onto the grass.

As the canine officer got off the man's back, he reached down and clipped the dog back onto its lead. He pulled it back as the other officers stepped in and lifted the man off the ground.

One of the officers got onto the radio, "Sarge, I think we've got your guy. He's a tall, skinny fella. We'll bring him out to the road. We're parked on Stirling Ave."

"Yeah, I'm almost there," grumbled the sergeant on the radio.

"Why did you sick the dog on me?" yelled the male, who was partly limping and partly dragging out of the backyard area.

"I warned ya, young fella," chastised the canine officer who was following the fresh catch. "When a police officer yells at you to stop, you stop, you understand?" he continued to berate the young man as they came out to the curb in front of 372 Stirling Avenue.

"There. Sit," Commanded one of the officers as he motioned to the young man to sit on the curb. The two constables searched the man and pulled a wallet and a set of keys from the man's pockets.

"Now, son, my dog was pretty intent on getting more of a piece of you, which means he smells drugs. Now you hand 'em over, or my dog will go find them, and your arm won't be your only bobo from tonight," said the canine officer, holding the dog close by his lead.

"I ain't got drugs, sir; I'm carryin' in the back of my pants." The man replied as the tears rolled down his grass and dirt-stained face. The officer who had directed the man to sit on the curb reached down, lifted the man by the handcuffs, and held them up high as he fumbled around the back of his pants and pulled out a small six-shot revolver.

"You got a permit to carry this, son?" the officer asked as he let the man fall back down, off balance, onto the curb.

"No, sir, I do not," the man replied as his head sank towards his chest.

"Why in God's name would you steal that car and then light another one on fire?" asked the other constable as the sergeant's cruiser pulled up and stopped. The sergeant exited his cruiser and looked around to ensure no people were staring from their houses at that late hour.

"I didn't steal no car, and I didn't light that car on fire, I swear." The man never looked up as he answered.

"You're saying it like you were there, son. Were you there when the car was lit on fire?" the constable asked.

The man looked up to the constable. "Yessir, but I didn't light it on fire. I was walking on the street, and suddenly, there was an explosion, and the car was on fire. I don't know how it happened, but I didn't do it." The man sunk his head back down as the looks of disbelief crossed the police officer's face at the ridiculous statement he had just made.

Sergeant James stood over the man as he asked the constable for his identification. He flipped through the man's wallet, pulled out a driver's license, and examined the picture.

"Head up," the sergeant growled as he used his long, hickory police baton under the man's chin to lift his head.

The sergeant's eyes closed as he breathed in a long breath of frustration, "this ain't George Cooper," he gritted through his teeth as he stared at the brown-haired young man.

"No, I ain't George Cooper, sir. I don't know who that is," the man pleaded as the hickory stick held his face up toward the sergeant.

"No, you ain't; pretty sure George Cooper is a made-up name anyways," Sergeant James replied before he pulled the baton away from the man and slid it back into the holder on his belt. He looked at the constables, who had confused looks on their faces.

"This ain't the guy I'm looking for, but book him all the same for starting that car fire," Sergeant James ordered.

"Good work," the sergeant mumbled as he returned to his cruiser.

He stopped with the door open and started pointing at the three officers standing there, "You, book this guy, you, get back looking for that other kid," and he pointed at the canine officer, "Good work, Fredricks, now I need you and your dog to keep in the area so we can get that kid from earlier. It will be a long night 'cause we ain't leaving until we find him."

Sergeant James got back into his car, and as he drove off down the street, the bustling constables could hear him screaming obscenities in the vehicle as his temper had gotten the best of him.

Chapter 47: Setting up at the CFH Telescope

After spending the first couple of days setting up their accommodations, Leonardo and Kevin felt more comfortable in their new surroundings. The routine at the Canada-France-Hawaii telescope was simple since Kevin's time with the telescope for the first three months was 2 a.m. to 5 a.m., six nights a week. Leonardo had been taking as many safety measures as possible while Kevin was sleeping or conducting his research at the telescope.

AI guided Leo along the way regarding placing several nano cameras and nano weapons. And today, Leo was taking a taxi back down the mountain to place two nano cams to establish an outer perimeter of defense. It would be the first time he had descended the volcano since he and Kevin arrived two days earlier.

Kevin was waking up from his late morning snooze, and he caught Leo heading for the apartment door with a briefcase in hand.

"Where you off to, Stan?" Kevin asked as he rubbed

his eyes, walking into the small living area.

"Headed down the mountain to meet with one of my contacts. I'll only be gone an hour or two at the most. I left you some eggs and toast on the stove. I figured you'd be hungry since you didn't eat after you got back this morning.

"Thanks, Stan, that's very nice of you," Kevin said as he stretched out his long arms and legs simultaneously; he resembled the figure of Gumby as he stretched. Kevin continued, "Are you always this considerate to the people you protect?"

Leonardo smiled, "No, not really, but I figured that since we will be staying together for the foreseeable future, I'd help out wherever I could so you could just focus on your research." Leo paused and scratched his head before he spoke again, "I'd like to come with you tonight to see the telescope in action if that's okay with you. I find it very interesting."

"Absolutely, Stan, I welcome the company, and I can show you around. Making sure the telescope was safe for us on the first day was probably not a great self-guided tour," Kevin smiled at Leo.

"No, I'd like a proper tour if you have the time. I don't want to interfere," Leonardo said as he swung his briefcase back and forth.

"Okay, Stan, get going," Kevin laughed, noticing Leonardo's wish to get going out to the cab that had pulled up moments earlier. Leo laughed as he turned and walked out their front door, across the skinny porch, and down onto the ground where the soil revealed the path most traveled over the years.

"See you later," Leo shouted as the screen door to the apartment boomed closed. He smiled as he walked out to the yellow cab waiting there and greeted the driver as he got into the back seat.

"Where to, ho aloha?" the smiling cabbie questioned as he backed up the car to turn around and head back down the mountain road.

"Headed to the airport to meet up with a friend," Leonardo responded as he was distractedly fumbling around in his briefcase.

"Sounds like a plan," the cabbie said as they started their twenty-two-kilometer descent down the mountain roadway.

"Oh! And I want to stop partway down the mountain and then near the bottom to get out and take a picture; it's so beautiful," Leonardo said.

"You got it," said the cabbie.

Leonardo was in the back seat watching AI's display inside the briefcase on his lap to better instruct the cabbie when to stop so he could get out and mount a nano camera on a tree at the side of the roadway. Leo had done all his research before leaving and knew that the predominant tree in the higher altitudes was Māmane, or Sophora chrysophylla, as AI listed the tree by both names; these trees grow to be over ten meters tall, ideal for camera placement.

Unfortunately, as Leo and Kevin had ascended the mountain two days prior, Leo was not able to have AI scan the vegetation to determine where the best placement was for a nano camera; however, AI approximated the two distances down the mountain road, which optimized their field of view for all traffic traveling up the mountain.

Leonardo prepared for the first stop by latching the

first nanobot cradle to his index finger; this nanobot was pre-programmed to latch onto whatever surface Leonardo touched his finger to. The cradle for each nanobot was essential since the human eye could not see so small an object without magnification, and the cradle served as a beacon to reflect any light to the eye. The nanobot cradle looked like a tiny speck of glitter catching the light just right to an unknowing person. Leonardo closed and locked his briefcase as he saw that the countdown display was within the two-hundred-fifty-meter buffer zone where he could deploy the camera.

"Sir, can we stop here where it's safe to stop? I'll be a minute while I get a picture." Leo asked.

The cabbie nodded and smiled at Leo in the rear-view mirror as he signaled to pull over to the right of the roadway. As the car slowed, Leo saw a tall, sprawling Māmane tree.

"Right here, please!" Leo pointed as he got ready for the car to stop.

Luckily, Leo didn't undo his seatbelt yet; as the car slowed more aggressively to accommodate his request, the force pushed him forward. Leo felt off balance as the car lurched to a stop since the road grade was steep. He got out of the cab, holding his briefcase, and walked back up the mountain a short distance; it wasn't long before he was puffing like an old steam train to catch his breath.

Leonardo was used to living in a high-altitude country like Switzerland, which had an average elevation of 1350 meters. Still, comparatively, the telescope he and Kevin had been staying at was at a height of 4200 meters. This waypoint chosen by AI was at least 2,500 meters in elevation, so Leo had to work even harder with the much thinner oxygen levels.

Leo got to the relatively large tree at the side of the roadway, and as he put his briefcase down beside him, he mimicked taking a photo with his back to the cabbie. The cabbie was paying no attention to Leonardo; he was listening to the CB radio and getting information on his next fare.

Leonardo walked a short distance off the road, reached up with his right hand, and placed the nanobot as high as his arms would reach. Hearing the single, audible 'beep' from the AI in his briefcase indicated that the nanobot was successfully attached to the tree and could reposition itself for the best possible view of the roadway.

Leo picked up his briefcase and walked back towards the cab. He was surprised how hard it was going down the steep grade as his quadriceps burned, slowing his body's speed as he balanced the steep hill.

At least this altitude will help me get in better shape, Leo thought as he reached the back door of the cab, got back inside, and fastened his seat belt.

"Thanks," Leo said as the cabbie put the car into gear, and the brakes squealed as they continued down the mountain.

Again, Leo opened the briefcase on his lap and examined the distance to the next waypoint that AI wanted the next nanobot deployed.

"What you got in the case, sir?" the cabbie asked as he examined Leonardo in the back seat.

"Oh, just some notes I'm reading for my meeting with my friend," Leo swallowed hard as his throat rejected his lie, "he's my friend, but I also need to talk business with him." Leo smiled back at the cabbie.

"Okay," the cabbie answered as he picked up the CB radio and joined the chatter over the airwaves.

Leonardo didn't pay any attention to the different language cadences and pronunciations broadcasting on the radio; he focused on the distance to the next waypoint. The countdown distance was still displaying eight kilometers as they wound their way down Mona Kea.

Leo thought about Bruce and hoped he was doing okay. Bruce had not communicated after his cryptic napkin message indicating that he would not be on the flight, and Leo was trying hard not to worry.

He hoped that the message he had programmed AI to transmit as their flight from LAX was getting up to cruising altitude, so to speak, didn't fall on deaf ears. Leo was thankful that he insisted on duplicating the AI unit so that Bruce could access many of its features. He looked out the window at the breathtaking scenery as they descended into a larger canopy of trees and took a deep breath.

Please be alright, Bruce, I can't do this without you, Leo pleaded.

Chapter 48: Getting out of Dodge - Again (03.12.1981 00:13 hrs)

Bruce got out of the car's back seat, slung his backpack onto his back, and straightened his ballcap as he stepped to the side of the road. The vehicle then continued and turned north.

"Thanks!" Bruce yelled as the car left. He looked around before stepping farther off the roadway and then talked to AI, who had been silent for hours.

"AI, where the hell am I?" Bruce asked.

"Your location is close to the intersection of Interstate 10 and Highway 177 in the small community of Desert Center, California," AI responded.

"How far to the airport?" Bruce asked as he looked around him at the vast open desert. The crickets were singing their hearts out, and the moonless night air was cool but dry. Bruce needed to find a place to rest before continuing.

"LAX airport is 305 kilometers from your location.

It would take approximately 3.25 hours by vehicle. That would translate to…"

"Yeah, yeah," Bruce cut off AI and started walking West on the side of the highway. He was thinking. He had been walking or hitchhiking for more than twenty-eight hours, hoping his trail would go dead for the police looking for him after his late-night escape through the desert.

While walking, he stopped to drink any reasonably clear, flowing water he could find to stay hydrated for his long trek; the nanobots in his bloodstream were no doubt killing multiple different types of pathogens from the contaminated water Bruce had consumed while on the run. He took out the small water bottle he had filled up hours earlier before being picked up by his most recent ride; it was empty.

He opened the bottle, tipped it up, and felt a couple of drops land on his tongue, but it wasn't enough moisture to swallow anything. He closed the bottle, put it back in the side pocket of his backpack, and continued walking.

"AI, continue scanning for water sources as I walk. Radius one kilometer," Bruce asked as he nodded to himself at his specifications; at this point, he would go a kilometer out of his way for a nice drink of water.

"Scanning," AI answered as Bruce continued up the side of the highway. He passed by a small trailer park on the North side of the road to limit his interactions with people, and none of the vehicles there were ideal to steal.

Bruce was thinking back to when Leo was setting up the temporary AI unit that Bruce was now using to his great advantage. He thought about Leonardo and Kevin, who had been in Hawaii for two days, while Bruce walked and thumbed across the state. He smiled at the irony that Leonardo, the leading scientist for the HERB mission, was now in the middle of the mission, and Bruce, the mission specialist, was now out of touch and wondering how the mission was going.

It wasn't going so well for Bruce, but he was confident he would get to LAX in one piece and get on a flight to meet up with Kevin and Leonardo. He had to; there was just no other option.

"AI, continue scanning for water but conserve energy wherever possible. I need you to be fully functional all the way to Hawaii, with time to spare."

"Copy," AI responded, "battery life will be acceptable for three more days, plus or minus 0.5 days, depending on usage."

"Ok, thank you." Bruce did not want AI talking needlessly. However, as he walked, he thought about making a few modifications later to benefit a user of the AI unit because its conversation abilities were dry.

Bruce continued walking westward for an hour before he could see the outline of a bridge in the distance; that meant two things: shelter and water. He had wondered why AI hadn't told him there was water up ahead and thought maybe he wasn't within a kilometer of the bridge for the AI to announce the water up early.

Bruce kept walking and finally reached the highway sign announcing the bridge up ahead. He waved his hand back and forth in front of the nanobot camera still perched on the peak of his ballcap, but there was no response.

Funny AI, Bruce thought. As Bruce got closer to the bridge, he veered farther off the roadway to access the shelter under the bridge without being seen. His fear was confirmed as he got to the embankment and the riverbed below; there was no water. The bridge was there to accommodate the water flow when there was water. However, the dry weather and hot Sun had dried and baked the riverbed into a cracked and sharded appearance. It looked like the riverbed was glass that the Sun had smashed into a million pieces as it lay on the ground unprotected.

Bruce was drained. He had barely eaten since the previous morning, except for the piece of beef jerky the last driver had given him. His mouth felt like the husk of a riverbed he was examining. He decided to get a few hours of rest before continuing his journey.

He took out the nanobot-laced fabric he had rolled up tight in his backpack, sat down on the cement rise directly under the bridge, and unfolded the fabric to cover himself up. The material's properties would automatically adjust to keep his body heat reflected inwards, keeping him warm as he snoozed and minimizing dehydration by not letting him perspire by getting too hot.

Bruce laid down with his arm hooked into the backpack strap and placed his ballcap higher up on the cement rise so that AI could watch as he tried to get a couple of hours of sleep.

"AI, keep a lookout for any humans, animals, or reptiles while I get some sleep." Bruce added in one more detail that he hoped would get him some sleep, "AI, radius of safe zone ten meters, and do not include vehicle traffic going over the bridge."

"Copy," was AI's response.

Bruce felt like he had barely shut his eyes when he awoke with a start, unsure of where he was. His eyes darted around as he sat bolt upright and took in the dimly lit surroundings.

The memory of the past thirty hours inundated Bruce as he relaxed slightly and started rolling up the nano fabric that had kept him warm for his almost one-and-a-half-hour nap.

Bruce smacked his lips together and felt them with his fingers; they were cracked and dry like everything else in his environment. He got his bearings and walked out from under the bridge and back up the same path he had taken toward the riverbed earlier in the evening.

Bruce looked at his watch; it was 2:43 a.m. as he

started walking west down the side of the highway. He was going to have to find water soon.

"AI, continue scanning for water with the same parameters as earlier today," Bruce asked as he cleared his aching throat.

"Copy," AI responded.

Bruce admired the waning crescent Moon that had come up in the East, pointing to the Sun, which would follow in a few hours. He wouldn't need to have an AI scan for people or animals since Bruce's low light vision was excellent; that would also save battery life for his precious AI unit. He was getting into stride as he saw cars come from the horizon and go over the opposite horizon as he walked.

He was walking for half an hour before noticing a small house off the side of the highway about five hundred meters away. There was a single light on inside the wooden shack-type house.

As he got closer to the house, Bruce activated AI. "AI, scan the residence ahead and the property around the house. I want to know how many people are there."

"Scanning. There is one person, and one canine

showing on the thermal scan. There are no other humans or animals in your vicinity."

As the distance to the house diminished, Bruce switched AI to silent mode and gave one last direction before he made contact with the owner of the house, "AI, have NDI ready to launch on my command word 'help,' or if my vitals show high levels of stress hormones," Bruce directed as his distance closed in on the tiny house. *Please be a nice person,* Bruce thought as he got to the door and knocked, and immediately stepped to the side of the doorway as he waited, lest he be shot through the door if the owner did not take kindly to people walking up to their house unannounced.

"Who is it?" yelled an old, frail-sounding voice from the other side of the door.

Bruce could hear the person shuffling on the old floorboards towards the door when he answered. "I've been walking all night, wondering if I could have some water?" Bruce asked in the most likable voice he could muster, given that his throat was so dry.

The door creaked open a few inches, and Bruce

could see a little older man standing on the other side of the door. The man spoke again, "You best not be trickin' me, boy. I'll give you water, but if you try to steal from me, I'm gonna put a bullet in ya," the man said as he peered through the crack in the doorway.

"You have my word, sir," Bruce answered, "if I could just have some water, I'll be out of your hair in a jiffy," Bruce finished, giving his best dehydrated smile. He could feel his lips cracking as he forced a smile.

The door opened slowly, and the man turned and shuffled back to his chair. The dog was geriatric; it hadn't moved a muscle except to look at Bruce as he stepped into the house.

"Thank you so much, sir," Bruce said as he walked in and stepped into a more open area to the right of the door. The house consisted of a single bedroom at the back, a bathroom, a tiny kitchen, now behind Bruce, and the living room where the elderly man was back sitting in his rocking chair.

"Help yourself; the water from the tap is pretty brown with this hot weather, but I've been drinking the water for fifty years, and I ain't ever had any problems with it," he chuckled as he rocked.

Bruce turned and went straight to the little sink behind him, turned on the water tap, and started drinking as heavily as the flow of water would allow him; he gulped and gulped the water until he had to stop briefly, hanging his head in the sink to take a few breaths before he went back to drink more. The water tasted silty with a metallic aftertaste, but Bruce didn't care. *Getting that waterborne parasite and disease nanobot injection is feeling pretty good right now,* Bruce thought as he continued to tank up on the tap water.

"You was thirsty," the man commented, "drink your fill; I'm headed to bed shortly, and I'm afraid you can't stay the night here."

Bruce stopped his guzzling briefly to respond, "Yes sir, I'll be but a minute, and I'll fill up my water bottle, and I'll be out of here." Bruce went back to drinking until he felt like his stomach might burst. Then he took out his water bottle, filled it up to the top, and put it back in the side pocket of his backpack.

He took off his ballcap and bent down, immersing his head in the water, letting the water run over the back of his head down to the front before shutting off the tap and wiping his head with his dirty shirt. Bruce straightened up, put his ballcap back on, and turned towards the man sitting quietly in his chair.

"Thank you, sir. I appreciate your generosity," Bruce said as he tipped his hat to the man and walked to the doorway.

"Good luck, Sonny," the man responded as he got up and held his balance with his hand on the top of the rocking chair.

Bruce nodded as he turned and left the house and closed the door behind him.

He felt so much better as he walked back towards the highway in the dim light of the Moon; the water in his belly sloshed as he walked. He smiled at the simplicity of how happy he felt just being given the gift of water.

When Bruce was one hundred meters from the house, he started quizzing AI and formulating a plan to get to LAX with the least possible risk.

"AI, how far to the next town?"

"25.7 kilometers to Chiriaco Summit, California, West of your location on this roadway," AI answered.

"Please tell me there is a Greyhound bus station there," Bruce pleaded.

"Unknown," said AI.

"I guess you can't know everything since there is no digital information network for you to connect to for ten years." Bruce wondered what AI would respond to that as he picked up the walking pace and checked the time on his watch; a twenty-five-kilometer walk meant that he could conservatively get to the next town by 7:30 a.m. and hope they had a morning bus going through the town going West.

"There are no digital signals in this place," AI commented.

"No, there are not," Bruce chuckled, "It is a luxury that has allowed me to get this far. Now, silent mode for you, AI. I need you to conserve power," Bruce finished. He didn't want to waste precious scanning time by letting AI pontificate why a lack of digital technology at this time was so convenient for the time travelers who were desperately trying to reunite.

Bruce thought about the day long ago when he was traveling with Leonardo. Leo was bursting with joy at not having any microphones, cameras, or sensors around him except for the ones they possessed. He could now relate, as a lack of technology had allowed Bruce to escape the police without a single photo, video, or scan that would enable them to find him.

At best, they would have a grainy photo from the police station or get a partial fingerprint from the car, but given the primitive forensic ability of the police in 1981, Bruce was not concerned at all. The police would have a set of descriptors of his facial appearance, height, weight, and what he was wearing that they might send to neighboring police precincts. Still, Bruce was counting on being in the clear once he crossed into California during his last car ride a few hours ago.

As he walked, Bruce had loads of time to think; he was hoping to catch a highway bus from the next town and jump busses to the airport, but he also was birthing plans to have as a backup should there not be any transportation from the next town. Bruce had dug a small hole near the bridge he had slept under earlier in the night and buried the identification card with "George Cooper"; he would never use that name again.

He was fortunate to have an American passport with a different name hidden under the bottom support of his backpack. However, it was not luck since Bruce had the "George Cooper" throw-away identification specifically for any unknown occurrence.

Bruce laughed to himself at the set of unexpected circumstances that had gotten him to this moonlit highway in California as the words of Frankie, one of his mission preparation colleagues, came to him, *"Bruce, we plan for the unexpected, even if it seems like your primary plan is a surety for success; why? Because we never expect the unexpected."* Bruce and Frankie would say the last part together and laugh as it was being hammered into Bruce's mind.

He made a mental note to get down on his knees and thank Leonardo Hoffman for saving the mission; if Leo hadn't forced the temporary AI unit on Bruce, he wasn't sure where he'd be. It would be like flying a plane in darkness without lights, flight controls, or radar. As Bruce thought about the metaphor of flying in the dark, it reminded him of something AI had said to Leonardo as they were preparing to extract Kevin Nault, "When the Moon is appropriately positioned, I can reflect a signal of the lunar laser ranging retroreflector array installed on July 21, 1969. That would translate to a signal broadcast distance of approximately 4,000 km, depending on the Moon's location in its orbit around the Earth. "

Bruce looked up to the Moon overhead and wondered how far Hawaii was from California to the island of Hawaii. He thought about it for several minutes, and when the suspense was killing him, he had to ask AI, "AI, how far would a signal have to travel from here to get to the island of Hawaii, and is the moon in the correct position for me to bounce a signal off the array on the Moon?" Bruce asked.

He listened with bated breath as he thought of how much a relief it would be to Leonardo to get a message from Bruce telling him that he was safe. Bruce didn't have to wait long, "The distance from the island of Hawaii to this location is approximately 3885 kilometers; however, the position of the Moon is not ideal for the signal to be received with high certainty. We would require the Moon to be…" Bruce cut off AI.

"AI, what is the probability of successfully sending a message to Leonardo from this location?"

"Probability is 0.76," AI answered.

"Good enough for me," Bruce answered with a smile at the chance of getting a message to Leonardo.

"AI, prepare a message to reflect off the Moon's array to the island of Hawaii. Encode wave burst with my location, date, and time, and encode this message in text only in five Gigahertz bandwidth; send when the message is complete." Bruce thought momentarily; he wanted the message to be simple and not cause Leo any extra worry.

"Hi Leo, I got a little sidetracked, but I am back on course to the airport. I hope to be with you within the next 48-72 hours. Best wishes, Bruce." Bruce laughed out loud as AI sent the message; Leonardo was not a 'best wishes' kind of person or a 'regards' person. Leo said what he meant and was not a fan of impersonal messages. He thought that Leonardo would get a little chuckle from the 'best wishes.'

"Message sent," AI confirmed for Bruce.

I hope you get that message. Bruce looked up to the Moon, willing his message to successfully reflect off the Moon's array and be picked up by Leonardo's AI unit which continually monitored the 5GHz bandwidth. In 1981, humanity did not have the technology to use high-energy bandwidth, so either Leonardo would get the message or it would be lost forever. Since Bruce's AI was not given the ability to receive messages, he would be in the dark, not knowing anything about Leonardo or Kevin until he was there in person.

Chapter 49: Watching the Stars

Leonardo had returned to the researcher's residence before dinner as the sun dipped from the volcano-top sky. He sat down while Kevin was heating a couple of cans of stew on the stove. Leonardo was still having bad headaches at the high altitude and was feeling dizzy when the cab dropped him off; Kevin, luckily, was there to greet him and ushered him inside to sit down.

"It'll take some time to get used to the altitude, Stan," Kevin remarked as he returned to the stove and slowly stirred the two portions of stew in the pot.

"Yes, it might be headaches from an old injury," Leo said as he gently rubbed the side of his face that now bore a considerable scar. *It still ached.*

"Well, regardless if it's the injury or you're just having trouble adapting to this altitude, you just have to ease into it. If it's not better by the morning, you should go down the mountain to get sea-level air. It must have felt nice today to get full air."

"It was quite nice, I will admit. I need to get used to this altitude," Leo groaned, "I can't just leave you here anymore; I have a job to do." Leo continued.

"Yes, you have to protect me," Kevin rolled his eyes.

"What?" Leo asked, straightening his head at the sarcasm.

Kevin laughed as they looked at each other, and Leo smiled at the sheer comedy of their situation. *Kevin, you have no idea,* Leo thought as they smiled at each other.

"Seriously, Stan," Kevin was still smiling, "I don't doubt your defensive skills, but you're not in as good shape as I would have guessed for an FBI agent."

"What?" Leo sat bolt upright, but his hand automatically went to the side of his head as the throbbing continued. He caught himself smiling and spoke again, "Well, I go from zero to one hundred for a few minutes or hours, and then it's sit, sit, sit again until something else happens, you know?"

Kevin was studying Leo's face, "Yeah, I understand. I'm just teasing you, but seriously, where is your gun? I haven't seen a hint of a gun this whole time, Stan, not even when we went through security for the flight. Is it a secret gun?" Kevin was being sarcastic again.

Leo's face became serious, "Kevin, you're just going to have to trust me. You need me here, and if something happens, I will protect you. Not everything is as it seems."

Leo's attempt to be mysterious was lost on Kevin, "Okay, mister secret weapon, I trust you, but I'm still not understanding how Bruce and I got dragged into this terrorism thing. Bruce especially! He's just a kid. A brilliant kid," Kevin lamented Bruce Hayden's genius at such a young age. He continued, "He doesn't deserve this."

"Well," Leo answered, "From where I'm looking from the outside, I think this is a win-win for both of you. You get to spend an extended time researching at this brand-new telescope, and Bruce learns from you. Do you think he will lose out in any way by learning from someone like you?" Leo asked as he was sitting back in his chair again.

"No, I suppose not. I know Bruce and I can do great things together. I'm sure of it." Kevin was still slowly stirring the stew as he spoke.

That's the understatement of the millennium, Leo thought.

Leo needed to change the subject quickly. He felt uncomfortable with how relaxed he was around Kevin and did not want to slip up because of his comfort level.

"Uh, do you think you'd have time to show me around the telescope facility tonight?" Leo asked.

"Yes, of course," Kevin responded as he shut off the stove element, "I am starting my research observations tonight, so maybe we can go in early before my telescope time starts, and I'll give you the tour, but then I have to get to work; I don't want to miss one minute of my telescope time," Kevin pulled the pot of stew off the stove and started scooping it into the two plain white bowls that were sitting there at the ready.

"Absolutely, that sounds wonderful," Leo responded, concealing his excitement. Thrilled was not a strong enough word to describe how Leonardo felt about witnessing history in the making. To Leonardo, doctor

Kevin Nault had gone from a mission-critical person in the past that he had studied and admired, to a person that he was now staying with *while* they were leading up to the monumental discovery of the asteroid and then building landing pods and creating new technologies.

Leonardo was in awe of what circumstances had brought him to the front lines of the mission. He found himself being grateful, even though it meant that he had to get pistol-whipped, shot, baked in the desert heat, hunted by a maniac, and exhausted in the high-altitude oxygen to get there.

Leonardo's daydream was interrupted by Kevin, who plopped down the steaming bowl of stew on the small, rickety coffee table in front of Leo.

"Eat up, Stan. You need more energy for your body to function properly at this altitude," Kevin smiled as he backed up and sat in a wooden chair across from Leonardo with his stew, "I'm just glad you're not a smoker," he said as he blew on the spoonful of stew that was steaming heavily.

The steam extended into a stream, propelled across

the bowl and into the room. "You'd have a hell of a time catching your breath at this altitude if you were a smoker." Kevin finally tested the stew and took a tentative mouthful, blowing air out of his open mouth to help lower the temperature of the tasty food.

"No, smoking is so horrible for you; I've never tried it," Leo added as he watched and waited for his stew to cool down. "It should be illegal, you know." Leo picked up the bowl and rested it on his legs while alternating hands to support the balance of the bowl because the bowl was so hot. In the future, smoking would be deemed a scourge to many country's health care systems, and after governments had successfully sued multiple tobacco companies for chronic health problems, tobacco for smoking was a thing of the past.

"I think so, too," Kevin answered. "Don't get me started about the smoking section of the plane; that was horrible!"

"I know, right?" Leo laughed, "They should have smoking and non-smoking flights because the smoking section infects the whole plane!"

Kevin chuckled and nodded in agreement as his mouth was full of a fresh spoon of stew.

"You're going to make me spit out my stew, Stan," Kevin laughed as he finished swallowing the mouthful he had just taken.

Leonardo's stomach finally got the best of him, and he tried a spoonful of the stew. He was surprised that it wasn't as hot as he thought.

Kevin said, "Off topic, but did you know that water boils at around eighty-six degrees Celsius at this high altitude? I did the calculation while I was heating this gourmet dinner."

"Mmmm," Leo answered as he was too late to abandon his mouthful of stew.

After a few moments, Leo was able to answer correctly. "I thought the stew wasn't as hot as it looked. Turns out it's because the boiling stew is fourteen degrees cooler than sea level."

"That's some good math, Stan," Kevin replied, "I'll make a scientist out of you yet."

"Ha!" Leo laughed a little louder than he had anticipated. "You can try."

The two men silently ate the rest of their stew as Leo struggled to return to character. *I'm Stan, and I'm an FBI agent. Not a scientist,* Leo thought. *I'm Stan, and I'm an FBI agent.*

Leonardo was again surprised at how disarming Kevin was; he had to focus so he didn't slip up. Leonardo thought about Bruce and hoped he was okay. He never thought being alone with Kevin Nault would be so challenging to stay in character; he thought airport security would be his biggest challenge when he realized that Bruce would not be there to assist.

Now that he had the three nano cams deployed: one outside the airport, one halfway up the volcano at the side of the roadway, and one at the volcano's summit, Leo felt more in control and safer. All three nano cams were weaponized, all with NDIs and lethal options; Shadow would have difficulty getting from the Hawaii airport to the summit of the volcano without being outed by the technology.

Shadow already demonstrated that he possessed lethal nano weapons. Now Leonardo and AI ensured that Shadow could not get past the airport without first having his 22nd-century technology detected by the nano cams.

Even when in a dormant state, the nano cameras and nano weapons emitted a faint but detectable signature if the detector was fine-tuned to look for them. Leonardo ensured that no nanotechnology would step out into the sea-level air at Honolulu airport without being seen; his knowledge of nanotechnology was not surpassed by anyone in 1981, even Bruce Hayden.

Leonardo had postulated that Shadow was trained as a disruptive operative, so it was unlikely that he had an intimate knowledge of nanotechnology other than how to deploy and set parameters as he had done at the attack on the Steward observatory.

Leo looked over and saw Kevin rifling through his briefcase, taking papers out, and putting them on the table.

Kevin glanced up and saw Leo looking at him, "It's a good thing we astronomers use the same computer programs and search algorithms in our research. All of my research files and most of my notes are still back in Arizona," Kevin mumbled as he pulled out a magazine and dropped it on the coffee table between him and Leo. Leo looked down at the magazine that had fallen sideways on the table, and he turned his head sideways to read the cover.

"Physics and Astronomy, July 1981," Leo read aloud as he again tried to silence his excitement at seeing the old paper document, which was new and fresh in science. Leo reached down and looked at Kevin as he slid the magazine closer to himself, "Do you mind if I?"

"Ha! Sure, Stan, a little light reading for you; fill your boots," Kevin laughed as Leo picked up the magazine and examined the photo of Doctor Richard Feynman. A pioneer of quantum computing, the first to postulate about nanotechnology and advance the theory of quantum electrodynamics, was pictured on the magazine's cover. Leonardo marveled at seeing an actual photo of the decorated scientist; Feynman had accomplished all the significant scientific milestones of his life by 1981 at the age of 61 but would live another seven years before he died in 1988.

Kevin spoke again as he saw Leo studying the magazine cover before he opened it to the first page: "You know, Bruce became involved in my research because of that magazine."

"Really? How?" Leo asked, unaware of the magazine's significance to Bruce and Kevin's scientific relationship.

Kevin closed his briefcase and set down some neatly piled papers on top of his briefcase before he spoke. "Well, I will never forget the day I met Bruce Hayden. He walked into the university and demanded he talk to me about joining the Physics and Astronomy program. Bruce intrigued me, so I brought him to my office and saw that magazine on my desk. I saw it sitting there, and I thought of asking this kid what he thought about quantum computing, which has been around in theory for decades.

I don't know why I asked him that because any other student wouldn't know the first thing about it. This young man told me about an application of quantum computing to computer simulation models like he knew it already existed, and that intrigued me to my very core. I'm sure it sounds dumb to you, Stan, but to me, there is something to

Bruce Hayden that enthralls me, like he is this undiscovered gift to the fields of physics and astronomy."

Leo was looking down at his shoe dangling over the top of his opposite knee. He was again shocked at the scientist's ability to judge character and to know on some level that Bruce Hayden was an essential part of his research. Leonardo wondered if the two previous missions in which Bruce was unsuccessful somehow impacted Kevin's ability to see great potential in Bruce. *A thought experiment for another time*, he thought.

Leo realized he needed to answer Kevin, "Uh, I don't think it's dumb, Kevin; I guess I just can't identify with what you're saying because I don't have your expertise in that field," Leonardo shrugged as he lied.

"But I'm glad we have him coming to stay here to help with your research and to keep him safe." Leo felt relief at being able to make a true statement.

Leonardo stood up and dropped the magazine back onto the table, "you mind if I try to read that later?" he asked.

"I'll leave it there for you," Kevin smiled as he

looked over the papers on his lap.

"I think I'll head out for a little walk to help get acclimatized," Leo said as he shuffled over to the door and outside into the crisp, cool air.

Kevin was already absorbed in the handwritten notes he was reading.

Leonardo walked the small loop around the residence building and examined the alien-looking surroundings; there was no grass, no trees, and no vegetation of any kind. The ground was a barren reddish-colored soil. *It must be this darker color either because of the Sun's rays at this altitude, or because of the low oxygen, or both,* Leo thought as he strolled around and around the small residence.

Four apartment units attached formed a square, each opening to a different side of the residence. The giant telescope building was one hundred meters from residence and stood up from the volcano's summit into the dark blue sky. The view from that altitude was breathtaking, although it was so high that the rest of the island was obscured by clouds that floated around the volcano like giant puffy cotton balls.

An advantage of the telescope at this altitude was that it was too high for cloud cover to obscure its view. There was little moisture in the air, which meant that the sky was crystal clear every night of the year. It was one of the best locations on Earth to view the heavens for nine more years, until 1990, when NASA was due to launch the Hubble Space Telescope.

The Hubble space telescope paved the way for science to continue its relentless march to discover the universe's secrets. But after some exorbitantly expensive and crippling malfunctions, newer generations of space telescopes were conceived and launched.

In 2021, the James Webb telescope was the first telescope launched from Earth that orbited around the Sun instead of the Earth; this allowed the telescope to be far away from any interference from Earth, and it was positioned so that the Moon or the Earth would not ever block the telescope's infrared view of the universe. The James Webb telescope's life spanned more than thirty years before it was replaced by the Unity space telescope in 2053.

The Unity telescope also orbited the Sun; however, in addition to seeing in the infrared spectrum, Unity could view the high energy spectrum and visible light; many discoveries related to black holes were achieved through the Unity telescope. These discoveries related to special relativity theory by studying black holes ushered humankind into the fusion energy era, which started less than a decade later. It turned out that studying black holes and how energy, light, and matter behaved while being subjected to extreme forces was the key to achieving fusion technology on a large enough scale to satisfy humanity's energy needs.

Doctor Leonardo Hoffman's life experience on Earth was from his birth in 2090 until the third HERB mission in 2135; after his unexpected travel back to 1981, he experienced a foreign time. His only knowledge of daily life in 1981 was from Bruce's briefings and debriefings before and after his failed first and second missions. Now, he was experiencing 1981 firsthand, which was a completely different ballgame.

Leo walked around the residence several times, daydreaming.

He did not like the feeling of labored breathing, and even though he knew he wasn't in good physical shape, he needed to be at his best to ensure that the mission could proceed on its improvised path.

There were still ten months before the modified asteroid discovery, and Leonardo had to ensure that the discovery would happen, Bruce or no Bruce. The Canada-France-Hawaii telescope was a giant, more reliable telescope that didn't depend on the weather, so the discovery of the asteroid could be made months earlier than at the now-destroyed Steward telescope in Arizona.

AI calculated that Shadow would not be expecting that the asteroid could be detected earlier than in previous missions, which was a distinct advantage to Leo and Bruce. As Leo thought about Bruce and watched the Sun beautifully sinking into the cloud deck, he quickened his pace, remembering that he should check his AI unit to see if Bruce had sent a communication.

It was an understatement to say that Leonardo worried about Bruce; no communication for almost three days was most unusual. Leonardo spent most of his time trying to occupy his brain with something other than worrying about Bruce.

The AI unit was aware it was in mixed company, so it would not give any audible or visual alert; Leo would have to check AI when he could sneak a peek inside his briefcase. The only exception Leo programmed into AI was that the signal would be sent directly to his watch if there was a security flag from any of the nano cams. As Leonardo turned the last corner of the residence building and was nearing the door to the apartment, he remembered the physics magazine Kevin left on the little coffee table for him to read.

I'll check the AI to make sure there was no message from Bruce, then I'll immerse myself in that magazine until it's time to go and see the telescope with Kevin, he thought.

Leonardo quietly stepped inside the front door to the apartment to see if Kevin was catching a few winks before his telescope time later in the night. He immediately saw Kevin standing in the kitchen area, immersed in some research papers he held. He paced back and forth in the small kitchen sitting area, engaged in his brilliant thoughts.

"Ah! You're back. How was it?" Kevin asked, his

concentration broken by the sight of a red-faced Leonardo coming into the front door. It was partly from the exertion of the walk and the nippy temperature outside, as the Sun was getting low in the sky.

"Good, good," Leo responded as he walked in, grabbed the magazine off the coffee table, and sat in the chair. "A little light reading, so don't let me interrupt you, Kevin," Leonardo said.

Kevin smiled at Leo briefly and was almost immediately re-immersed in his papers as he began to pace slowly back and forth across the small living space.

As Kevin read through his papers, the long fingers on his left hand slowly moved back and forth in an almost wave-like way, as if he were massaging the thoughts as they traveled through his mind.

Leo was pretending to read the magazine, and even though he desperately wanted to read the articles in it, he was more intrigued to watch Kevin in his natural state. Leonardo had heard many different stories from Bruce about the thought experiments he had taken part in with Kevin, and Leo wished he didn't have to pretend to be a cop instead of a scientist; it was any scientist's dream to be able actually to talk to and get to know their idols.

Unfortunately, most scientist's inspirations were men and women of the past who no longer roamed the Earth. Still, their theories and ideas continued to be examined, proved, or disproved by their future counterparts. Leo was now living in the same apartment as Doctor Kevin Nault, who would discover the menacing asteroid and help set events in motion that could save the human race from losing their home planet. Leo pretended to be reading, but whenever he didn't think that Kevin could catch him, he was studying the scientist.

I have to make a point of keeping a daily log of my interactions with Kevin. I want to remember every single thing happening right before me, Leo mused. It would prove difficult since Leo did not have the privacy to make a video log when Kevin was in the apartment. Having a near-perfect memory was a gift that Leonardo did not take for granted.

Making logs of ideas and thoughts was second nature to the scientist, and it was more a habit than remembering details of events. Leo would have to record all his logs during Kevin's telescope time when he was alone since using the AI anywhere else meant he could be overheard or seen. It was out of the question; Leo didn't want to make the revised mission any more difficult than it was already.

Time flew by as both men were ensconced in their own ways, Leo reading the ancient physics magazine and Kevin reviewing his notes, preparing for his precious telescope research time.

Before long, it was 1:30 a.m., and Kevin put his papers back into his briefcase.

He looked over at Leonardo, whose eyes were closed.

"Uh, Stan, do you still want to come for a tour before I start my research time?" Kevin asked quietly.

"Yes, yes, I certainly do," Leo responded immediately; his eyes were closed as he visualized what he had been reading. Leo got up, dropped the magazine onto the coffee table, and smiled at Kevin. "I'm ready when you're ready," Leo motioned to the doorway.

"You're not at all what I expected, Stan," Kevin said

as he grabbed his briefcase by the handle and put on the grey button-up sweater that he had been given upon his arrival to the cold volcano top environment.

"For a detective, that is," Kevin added, smiling.

Leo smiled back and shrugged, "I just have lots of interests," he added. Leo had a strange feeling that somehow Kevin knew that he was a complete fraud of an FBI agent and was playing along with the lie.

The two men walked silently in the crisp, dark air to the telescope building. When they reached the entrance, Kevin began talking about the telescope.

"So, this telescope is not quite two years old, and it's an optical and infrared reflecting telescope with a mirror diameter of 3.6 meters. That means it can see in the optical spectrum like we do and can see infrared light, a longer wavelength that humans can't see. My research will use optical light measurements and search for tiny variations in the starlight to detect large asteroids that are otherwise invisible to us."

Kevin and Leo walked through the front double door into the telescope building. Once inside the circular base of the building, they could see up into the area where the telescope was mounted on the second floor. The first-floor outer circle had a small kitchenette and some small offices. Then, there was a curved staircase up to the second floor, which was recessed slightly to allow the open concept of the building.

"Wow," Leo gushed as he took in the view of the old telescope. He was shaking his head at the sight and felt goosebumps rise on his arms and neck.

He would never have believed that he would be able to experience history in this way. Virtual reality technology in the 2130s enabled people to experience things as if they were there in person.

Sights, smells, touch, and taste; all the senses were satisfied with the technology, but Leo *was actually experiencing* these things in person, which made an indescribable difference.

He was in awe and trying hard to hide his inner scientist, who was excitedly jumping up and down.

"I know, right?" Kevin whispered as he smiled at Leo, and they walked past the small offices and kitchenette and straight to the curved open staircase that mirrored the outer curvature of the building. They walked up the stairs and entered the telescope area, which had most of its open space along the back half of the circle where the stairs came up to that level. It was surreal for Leonardo; he had only seen this telescope in a science history book and could now see it in person.

The two men walked slowly around the open area behind the telescope control area that had several computers on long desks. Three people were working at the computers, and as they saw Kevin and Leo, they glanced at their watches as they tried to finish what they were doing in their allotted time. They stopped walking and stood behind the railing, looking up at the telescope in the foreground of the amazingly bright stars that night. Watching the stars together, Leo could hear Kevin sigh as he looked through the large rectangular opening at the top of the building to the stars.

"This is my happy place, watching the stars," Kevin whispered.

"This is something else," Leo whispered back.

The two men stood in awe, watching the stars, as the other three scientists packed up their papers and ensured the data from their experiment was saved and exported correctly. Kevin was itching to get to his desk and start his observations, and Leo was anxious to let him as they inched closer to the discovery of the millennium.

Chapter 50: A Brief History of North America Part II

America, by the mid-2020s, was well on her way to sabotaging and bigotting her way to self-destruction. Social media and public outrage via the newly coined cancel culture allowed people to influence incompetent politicians and lawmakers into creating a society that was intent on forcing all their different beliefs on each other.

Abortion, sexuality, gender expression, and race were all topics that divided America in a fatal way, which culminated in the Civil War that burned America to the ground. She was blinded to the signs until it was too late.

Foreign powers contributed to the collapse of the Western world's superpower in their own way by inflaming and misdirecting news media, allowing the soon-to-be warring groups to reinforce their beliefs that they were right and the other groups were wrong. It was like pouring gas onto the already volatile and smoldering country.

In 2031, as America continued its death spiral, desperation drove the suffering country to attack its northern neighbor and ally. The attack on Canada was weak, divided, and uncoordinated; it proved unsuccessful as America's decades of self-abuse and in-fighting landed them in the societal gutter. It was a desperate move to try to occupy Canada and seize its natural resources: freshwater, minerals, and lumber.

By this time in America, protected forests, wetlands, and above and below-ground water reservoirs had been sold off to the highest bidder and exploited for financial gain. What remained of America in 2031 was a scorched, dry West, a heavily polluted East, and an utterly disengaged heartland. Many of the people of the West flocked to the ocean, virtually abandoning cities and rural areas in search of water; many large portable water cleansing units dotted the coastline and provided people and some animals with drinkable water.

However, flaring tempers and boiling frustration at the state of the country leeched chaos at every turn, and many water purifiers were destroyed because people were still pushing their own values on other people. America's nose was being cut off to spite her face.

Along the eastern coastline, many people traveled to the ocean as the great migration soldiered on; however, consistent precipitation allowed groups to remain away from the coast, where groups of people were jockeying to control the water purifiers. A complete dissolution of factories and businesses meant no other water purifiers were being built, at least not for the public.

The groups that remained inland in the east worked hard to keep the land clean, but heavy water and air pollution continually poisoned the ecosystems.

The American heartland suffered from extreme drought and polluted aquafers in the same period. Many people who chose to stay there were trying to restart the ecosystems and forming communities that grew environmentally sustainable food. Drilling to access new clean pockets of water had nearly emptied all known sizeable underground water reservoirs as deep as two kilometers.

Surviving by consuming plant-based diets and trading with other heartland areas allowed the people to live healthier than their cousins in the West and East. The communities cooperated to enable the people to live in harmony with fair trade of food, water, and materials for shelter. The people in the heartland avoided most of the conflict stemming from the political and resource-based fighting in the rest of the country.

With industry falling to its knees, all communication networks were cut off from society, except for people who had the equipment and the knowledge to operate it.

America had plunged into the first and only modern-day technological dark age that would last almost two decades before groups of people from the West, East, and the heartlands came together to cooperate and press forward to re-form a functional America. The resultant reborn country was better in all aspects and would forge the new America to be more balanced, inclusive, and environmentally conscious.

Although Canada did not fare as poorly as America at the same time, they, too, were trying to sort out many of the same societal issues that plagued their southern neighbor. One thing that set Canadians apart from Americans was their hearts. While America fought herself, Canada chose reconciliation, healing, and adaptation, with their hearts paving the way.

Natural resources had been protected, not squandered. Although Canada shared the longest land border in the world with the U.S.A., over the years and decades, they withdrew their resources from their greedy allies and became conscious environmentalists.

Air and water pollution streaming over the Canada-U.S.A. border drove Canada's largest populations to migrate North to more hospitable climate zones; global warming due to pollution had made the Southern parts of Canada dry and oppressively hot, whereas the more remote northern Canada became a desirable place to live.

The great northern migration in Canada spanned two decades, starting in the late 2020s; however, as the undeveloped parts of the North became settled, they heeded the environment's warning and created new settlements that worked in harmony with the environment. Canada survived the peak of the climate crisis only because of its immense geographic size and its willingness to adapt to change and protect the land. Indigenous people led the way for the people to live in harmony with nature, imparting 10,000 years of wisdom to help Canada survive the climate crisis.

No possible solution to help the environment was overlooked. Although none was the perfect solution, it was much better than what the country had focused on in the past. When the holy grail of energy sustainability was achieved in the 2060s, Canada was at the forefront, using the technology and helping poorer countries by sharing the technology and the knowledge to use it.

In time, Canada and America were rebuilt, better than before, and became global leaders again.

Chapter 51: Bruce on the Run (04.12.1981)

As Shadow left the police precinct and smelled the fresh Arizona evening air, Shadow felt confident in the move to expose Bruce to the 1980s law system. It was a bold but necessary move since all of Shadow's information and intelligence to locate Bruce Hayden had dried up like desert sand. The mission had taken a drastic turn from the now out-of-date data Shadow had studied repeatedly before accosting Doctor Hoffman in the HEPA lab.

While inside the police precinct, Shadow expertly deployed two nano cameras, and Bruce Hayden was on the run—not only from Shadow but also from the law.

Shadow had deployed one nano camera while walking through the constable's area of the precinct, and the other was deployed on the suit jacket of the lead detective who had taken the statement from Shadow.

One camera captured all the radio chatter and officer conversations from the precinct, and the other camera was perched on the glasses of one of the detectives.

It was privy to all the information the detective knew, in addition to the 22nd-century technology used by the nano camera.

Both groups of wavelengths, shorter and longer than visible light, were now being recorded from the detective's viewpoint. Shadow smiled as the finale of the brilliant plan was replayed in Shadow's mind:

"So, this fella's name is Bruce Hayden?" The detective tapped his index finger on the photo.

"He matches the description of a young man who stole a car and somehow escaped after coming in to give a statement about a crime he had witnessed the same day. Quite honestly, Lee, this young man left the old sergeant over at the twelfth precinct with mud on his face, and we'd very much like to find this guy for questioning."

"So would I," Lee responded. "I can't imagine what other reckless and dangerous things this Bruce Hayden has done. Imagine, that guy was at that observatory right when it exploded; it's like he was a thing possessed, and he must not have seen me because he was so intent on killing those men. And then he ran away. It looks really bad, doesn't it?"

"Yes," responded the detective, but his mind was still going over the details, looking at the still photo that this witness had brought in. "It is remarkable that you happened to snap this photo when you did."

"Yes, sir, I honestly don't know how I got him in that picture so clearly. Sheer coincidence, I guess, and when I had the film developed, I just had to come in and report this heinous crime."

The detective cut Shadow off, "So tell me, Lee, how did you come to be there as well? And you didn't present yourself to the investigator while they were doing the investigation. Why is that?"

"Well," Shadow shifted in the worn, padded chair, "I was scared that he would kill me if I came forward, but when I saw the man in the photo that I had taken, well, I just had to report it. I just stopped to take a picture of those nicely trimmed bushes with the observatory in the background, and this guy just walked into the picture. I started to get the camera ready to take another picture because I figured he had ruined it, and that's when the explosion happened."

The detective was mulling over the information on his pad of paper that had dramatically shifted the investigation from an accident into a homicide investigation.

He spoke without looking up from his paper. "Yes, and tell me about the other male you saw during that time frame." He tapped his pen lightly on the side margin of his notes where the word 'accomplice?' was circled several times.

"The other male was older and shorter, wearing glasses, and he was balding.

He was stalky but not fat. I saw him waiting at the edge of the clearing for the younger man when I got off the ground after the explosion. His face looked like he was yelling at the younger man, but my ears were ringing something fierce after the blast, sir."

The detective was writing furiously and nodded as Shadow gave the details. He looked up from his pad of paper after he circled something else on his paper, "What was the other man wearing?"

"I'd be guessing if I said anything specific, but I can

only assume he had a sleeve shirt and pants or shorts on," Shadow responded. *And show that the witness doesn't want to embellish the facts; it makes the witness more credible,* Lee thought, hiding a smile.

The detective watched Shadow's face as he spoke again, "Would you be willing to meet with our sketch artist so we can get a rough idea of what this other guy looks like?"

"Yes, I would be happy to help however I can," was Shadow's devious response.

The detective, who seemed happy with all the information he had jotted down on his pad of paper, collected all the pages from the interview and put them into order; it was not his first rodeo. He looked up and smiled as he put the papers together with a paperclip.

"So, Lee, can you stay with us today and see the sketch artist? I'll have to get a hold of him, but for information like this, we want you to meet with him right now.

"Yes, sir." Lee smiled as the nefarious plan had been executed perfectly.

Lee stood up and took the opportunity to deploy the second nano camera. Gently touching the detective on his left shoulder as they stood up, the nano camera was easily deployed. The nano camera had already been programmed to make its way onto the detective's glasses, where it watched and listened. Lee played the victim well; "I hope you find this man. I won't be able to sleep until I know he's caught."

The detective smiled at his victim as he reached out and firmly shook Shadow's hand.

"Yes, we will find him. It's what we do."

Shadow walked down the busy evening streets of Tucson, pleased that the advantage had now ebbed towards Shadow and away from Bruce Hayden. Shadow was smiling, listening to everything the unknowing detective was saying to the other investigators; it was beautiful.

Chapter 52: Scramble

Bruce had been walking for the rest of the night and early morning when he passed the nicely painted sign for the town of Chiriaco Summit; it had taken him more than an hour longer than he had anticipated due to the frequent moves away from the roadway as the nighttime traffic seemed to carry on without a break. Bruce did not feel comfortable hitchhiking anymore, since it would only take one person to report him to the police, whether they gave him a ride or not.

Chiriaco Summit was a small town whose population would grow to be a city in the future as suburbs of larger towns enveloped the landscape. In late 1981, the population was just over 25,000 people. Bruce walked down the side of the main street with his thumbs hooked under the shoulder straps of his backpack. The sun had peeked above the horizon, shining reddish light on the landscape. The morning traffic was picking up as more and more people started their days, and Bruce saw a sign for a gas station hanging in the distance as he soldiered down the side of the neatly kept street.

As Bruce approached the gas station, he watched the 'Pete's Gas' sign swing gently back and forth in the tender morning wind. Bruce got to the station, walked past the gas pumps, and a bell rang as he opened the door to the store. He did not see the little person behind the counter when she spoke,

"Can I help you, Sir?" Bruce looked around before seeing the little person step up onto a raised platform behind the oil and gas-stained counter. She had beautiful brown hair in a neat ponytail and a green shirt with the embroidered name 'Gail' with 'Pete's Gas' underneath. She started to straighten the packs of gum on the counter after speaking.

"Yes, ma'am, I'm hoping that there is a highway bus that comes through here, headed west." Bruce had thought long and hard about what his course of action would be if there were no bus, and it would mean either a lot more walking or dramatically increased risk via stealing a car.

"Yes, sir, but the bus only heads West in the evening on weekdays. If you're going on the weekend, mind you, the bus leaves here at 8 a.m. You can buy your ticket here, and the bus will be here about 7:55 p.m. tonight if you want to leave today."

Even though Bruce would have to wait all day, he was elated that he could catch the Greyhound bus, which meant that wherever the evening bus's last stop was, there had to be a connection to get to LAX. Bruce stepped up to the counter and smiled as he pulled a small bundle of money out of his pants pocket.

"Yes, ma'am, I'll take a ticket headed west for tonight, 8 p.m." Bruce smiled at Gail as she grabbed a ticket book from under the counter and filled out the ticket.

"How far west do you want to go?" Gail asked as she flipped the top paper away from the lower carbon copy layers to ensure she was writing hard enough.

"How far west can I go before transferring?" Bruce asked.

"This bus will take you as far as San Bernardino, and I believe you can transfer from there to go to Los Angeles. I'm just assuming you're going to Los Angeles," Gail smiled as she looked up from the half-written ticket.

"Yes, ma'am, good guess; I'm headed to Los Angeles. My dad can pick me up there, and then we're going to stay at the National Park North of Los Angeles. We both love nature," Bruce smiled a big toothy smile as he lied.

"Yes, okay, so I'll write this ticket to go to San Bernardino, and then you'll have to go into the terminal when you get there and purchase your ticket to go to Los Angeles." Gail wrote deliberately on the ticket to get the proper pressure for all the copies. "That'll be $37, sir."

Bruce unfolded the modest bundle of money he had ready in his hand. "There's twenty, thirty, thirty-five, thirty-six, thirty-seven," Bruce counted as he put the money on the counter in front of Gail. He put the remaining money back into his pocket.

"Name?" Gail looked over the counter at Bruce as her pen hovered over the left side of the ticket with four empty lines for name and address.

Bruce smiled, "Do I need ID?" As he pretended to fumble around in his pockets.

"Yes, sir, I don't make the rules, but you gotta have ID, and the bus driver will probably check too." Gail pursed her lips as she watched Bruce fumbling in his pockets.

Bruce finally pulled out his Arizona identification card and reluctantly put it on the counter beside the half-completed ticket.

"Bruce. Hayden." Gail said to herself as she printed the name on the proper line of the ticket, "Willcox, Arizona," she continued as she finished writing Bruce's information on the ticket.

Bruce swiftly pulled the identification off the counter and put it back in his pocket. He was uncomfortable giving his real name in this situation, but he had no choice since his throwaway ID had been compromised. Bruce watched as Gail ripped off the top copy of the ticket and slid it onto the counter; it was stopped by Bruce's hand before it sailed off the counter. She pulled out the bottom copy of the ticket and opened the cash register drawer as it let out a loud 'ding.'

She pulled the thirty-seven dollars off the counter, put it in the cash register, slid the copy of the ticket into the space between the cash holder and the bottom of the drawer, and closed the drawer. There was a hefty sounding 'click' as the drawer closed.

"Can I get you anything else, Bruce?" Gail asked after putting the ticket book back under the counter.

"No thanks, ma'am. I might come into the store later today before the bus to buy a snack or something, though," Bruce answered as he pretended to adjust the top flap on his backpack.

"Okay, good day," Gail responded as she stepped off the raised platform and scurried off to the side of the store that joined the garage.

Bruce took a deliberate look around the store before putting his backpack back on and walking out. *No security cameras. At least I have the day to think of how to get the copies of the tickets out of that cash register and from under the counter,* Bruce thought as he walked back out past the gas pumps and onto the sidewalk. He needed a quiet place to think and to consult AI; he didn't like how Gail's eyes sparkled at the young, handsome, red-headed man buying a bus ticket.

She would likely remember his name and certainly would remember what he looked like. Bruce walked down the side of the roadway towards the center of the town as small shops came into view and passed by. One shop caught Bruce's eye, and he stopped dead in his tracks. It was a pharmacy, and they had a small poster ad in the window for women's hair color. Bruce turned into the angled doorway of the shop and looked into the glass at the front of the building.

"Easy to use, takes only 90 minutes—a fresh, new look. Everything you need is included," Bruce read on the poster for the hair dye in the window. He lowered his ballcap over his eyes and walked into the store. Another bell dinged as he walked into the shop.

"What can I do for ya, young man?" An old grey-haired man said without missing a brush stroke as he swept the floor in front of the front counter.

"Uh, I'm just looking for some blister ointment, sir," Bruce stammered as he stepped into the store and pretended to be looking for something other than security cameras.

"Yes, follow me," the man leaned his broom against the counter and walked down an aisle, "here, this will do for you." The older man was short and stout and had a stereotypical work apron on with several items in the front pockets of the apron that poked the fabric out. He handed Bruce a small jar of greasy-looking ointment. "Anything else?" The man asked as Bruce grabbed the jar and examined the label.

"Hmm, yes, this will do. Oh, and sir, I see you have an ad for hair dye in the front window. My ma is having a heck of a time with grey hair, and I bet she'd like to try some of that." Bruce was looking from corner to corner of the ceiling as he talked to try and double-check for security cameras. No cameras.

"Yes, we have this hair dye that we know women will love. What color is your mom's hair naturally?" The man walked over to another couple of small aisles and stopped at a small display with a poster similar to the one in the front window.

Bruce squatted before the short display and looked at the available colors. "Brown," Bruce said as he grabbed a box from the collection that had a side strip of the color on the side. It was a lighter brown, so Bruce figured it would look more natural than black with his freckled face and pale skin.

"Alrighty, sir, I'll ring those through for you," the man held out his hand, and Bruce handed over the ointment and the hair dye box.

Several dings later, the cash register showed the total price for the two items as $5.50.

Bruce already had the cash in hand, dropped the $6.00 on the counter, and grabbed the two items off the counter, "good day, sir, keep the change," Bruce said as he walked out of the store and onto the sidewalk. A short walk West of the store, Bruce saw a road to his left with a heavily grown-over lot a short way down the side road. He looked around him as he got to the abandoned lot and stepped one giant step into a gap in the tall, thick bushes.

Once he was through the opening, a small opening close by was only about two meters wide and three meters long, surrounded by thick vegetation. A scorched area was in the middle of the opening, likely from other people using this little area to have a small fire. Garbage was in the fire, and a used needle lay on the side of the fire. He needed to get some rest and do some planning even before he got something to eat.

His stomach protested loudly as he put his backpack down and pushed some trash to the side to sit on a sizeable flat stone. It was angled nicely as he sat cross-legged and plopped his bag beside him.

"AI, I need security. Create a secure zone around this overgrown lot, and set geofence at five meters," Bruce said as AI returned to life.

"Secure zone, five-meter circumference," AI responded.

"AI, analyze footage from Pete's gas station and the pharmacy, with attention to my security. I need a way to get those copies of the bus ticket I bought."

Bruce took the two items out of the side pocket of his backpack and put them on the ground in front of where he was sitting on the flat rock. He opened the hair dye box and started to read the instructions when AI had completed its analysis. Bruce got a short vibration on his wrist from his watch as AI indicated it had completed the tasks.

"AI, go ahead with the report." Bruce put down the hair dye box to listen to what AI had seen.

"Pete's gas station has two employees working: one in the store and one in the garage. The employee in the store is female and short in stature, and she has a small fold-out knife hooked onto her belt on her right side. There is a sawed-off shotgun with a one-shot capacity hidden behind the counter, standing with some brooms off to the right of the counter from your viewpoint. The male in the garage is significantly larger than the female, and many tools in the garage area could be used as weapons. There are no security cameras nor any closed-circuit security of any kind. The cash register locks automatically when the drawer is closed, and a sequence of 4 numbers is pressed to access the cash register drawer.

Those numbers are 8-1-8-2."

"Haha, AI, I knew there was a reason I left you on the peak of my ballcap." Bruce laughed as he pictured Leonardo's face with the expression, *I told you so.*

AI continued after Bruce spoke, "The pharmacy has one closed-circuit security camera in the rear of the store above the pharmacy counter. It will have captured minimal footage of you from where you were in the pharmacy. There is also a full-length side-by-side shotgun behind the pharmacy counter standing in the rear corner; capacity two shots."

"Wow, two stores, two shotguns. Not surprising at all," Bruce rolled his eyes as he flipped the box of hair dye in his left hand.

"AI, give me some solutions to fabricate a new identification with the materials I have with me."

AI barely hesitated before answering as the petaflops of data were combed through and boiled down to a workable solution in fractions of a second.

"Your Bruce Hayden identification can be modified to show your new brown hair color and the alias Frank Hobden. It has the least changes that can be carried out with the laser sighting module on this AI unit."

Bruce smiled at the name change; it was a simple but realistic one that would have taken him much more time to figure out.

"AI, what parameters do you need to etch my photo identification in this space?" Bruce asked as he took out his photo identification card and looked at it from different angles. It had a thick lamination on it, and thankfully, identifications like it were low-tech and could easily be modified by AI. It would look entirely authentic to anyone who was examining it.

AI responded, "Place your ballcap on the flat stone beside you, with the photo identification standing upright, between twenty and thirty centimeters away."

Bruce thought momentarily as he took off his ballcap; he wanted to ensure he had thought of everything before modifying his ID card. He was still holding the ballcap when he spoke again, "AI, is there anything I am forgetting? I need this ID change to be seamless, including using the ID at the LAX airport."

"The identification card will bear the name Frank Hobden, an actual name in this part of the country. The name Bruce Hayden on your bus ticket can be modified by folding over the name multiple times, and then I can forge the name Frank Hobden onto the bus ticket with my laser. A distraction technique should allow you the time to obtain copies of the tickets in the cash register and under the counter at Pete's gas station."

"Great," Bruce responded, putting the ballcap down on the flat rock beside him.

Then he grabbed the hair dye box off the ground and put it about twenty centimeters in front of the ballcap.

Then he stood the photo I.D. up so it leaned against the box. He waited a moment to ensure the items were stable before he gave AI the go-ahead to modify the ID card with his new name. "AI, go with modification to my identification," Bruce said as he watched.

The alterations were completed in silence, and in the daylight, Bruce barely saw the light from the laser as it etched the name Frank Hobden in a way that completely concealed Bruce Hayden.

"Completed," AI chimed, and Bruce picked up the laminated card and looked at it from multiple angles to see how convincing the forgery was.

"Wow, nicely done," Bruce muttered as he shifted the card back and forth, viewing it from different angles. He fanned the card back and forth in the air several times before he put it into his wallet in the side pocket of his backpack.

"Now for the hair dye," Bruce said as he picked up the hair dye box off the rock and opened it. There were two bottles inside, along with the instructions. After skimming the instructions, Bruce removed his shirt and put it on the flat rock while he got into position with his head hanging over the small burn pile. He dabbed the contents of the first bottle over his hair so that the excess could drip off without staining his forehead.

Several applications were necessary to cover his red hair, but he could see his hair tone changing with each application as AI guided his hands to spread the dye out evenly. The plastic gloves from the kit were shockingly brown, and Bruce kept looking at his ballcap, which was still perched on the nearby rock, as he asked AI, "Is that good?"

Any person would have gotten annoyed at Bruce's constant questions about dye coverage and color, but AI, who was not capable of feeling annoyed, kept answering Bruce as he continued to apply the dye to his hair.

"More to your right, now down from that position. Apply more solution to the sides and back of your hair so that the hair showing while you are wearing your ballcap is sufficiently brown."

Bruce followed AI's instruction as he squeezed the last brown solution on the back of his head and then used the gloves to smear the color down to the bottom of his hairline.

"Lower…. Lower…to both sides…stop." AI said as Bruce gently pulled the rubber gloves away from his hair so he did not accidentally get any dye on his neck. He pulled off the first glove by turning it inside out, peeled the second glove off as it wrapped up the first glove, and tossed them into the fire pit.

"Now I just have to wait thirty minutes before I use the rinse solution," Bruce said as he stood up and paced around the open space of his temporary hideout. He kept raising his hands towards his head as he walked but resisted the temptation to touch his hair.

"So, AI, let's put together a more concrete plan for me to get those ticket copies out of the gas station before we finish modifying my copy of the ticket," Bruce said, but his mind was still thinking of Shadow, and what his next move would be. "AI, considering everything that has happened since Shadow got to 1981, what is the probability that the name Bruce Hayden will be burned in the next week? That Shadow character is very, very smart, tactically. And guaranteed he knows my name."

It took AI fractions of a second to analyze and extrapolate all the data of Shadow's tactics and known responses to Bruce and Leo's plans. Bruce squinted and held his breath while waiting for the suspected answer.

"The probability that the name Bruce Hayden will be unusable in the next week is 0.96. In the next month, that probability rises to 0.99."

Bruce's face contorted several ways as he digested the depressingly high probability that he could no longer use his name. It was a certainty and not a possibility. He took a couple of slow, deep breaths.

"AI, let's start with the plan to erase my name from that gas station's records and get me on that bus in one piece. Then we have to do some major damage control. If, or should I say, when my name gets burned, I need to still be able to work with Kevin in Hawaii, and he knows me as Bruce Hayden. We still have several hours before my bus arrives; let's brainstorm."

The human and artificial intelligence spent the next hours devising a plan to extract and destroy the two copies of Bruce's ticket at Pete's gas station that still bore his real name. Despite the slight chance that police investigators or, more likely, Shadow would go and ask the right questions at Pete's gas, Bruce had to make sure that no trail had a written copy of his name. More importantly, he had to devise a plan to completely change his identity for his research with Doctor Nault and the subsequent asteroid discovery-related fallout that would thrust them into a celebrity spotlight overnight.

The threads of the mission tapestry were being yanked from several different directions, and Bruce needed to figure out a way to tie off all the loose threads before the tapestry unraveled.

Chapter 53: Half-Truths

Kevin did not finish his night telescope time for the first time and stormed back to the dormitory building. It was chilly, even with a sweater at night, and Kevin walked with his arms crossed, more from anger than trying to keep warm.

He could see the distinct changing lights of a television screen illuminating the window from behind the curtains, and he thought it was fitting that Leonardo was still awake. Kevin was still reeling from his early morning surprise; *Bruce Hayden was wanted for two murders and was wanted for questioning in two other murders in Willcox*. Kevin immediately recognized the photo and sketch of Bruce displayed on the news network while they were giving details of the arrest warrants.

Kevin's head was swimming as he barged into the apartment, and Leonardo jumped up with a startle. He had been dozing on the chair with the television on.

"Hi Kevin, uh, how was your telescope time?" Leo stuttered as he stood up and rubbed the side of his face, which had a red palm print from his impromptu nap. He looked at his watch quizzically.

"Sit down, Stan!" Kevin shouted as he sat across from Leonardo.

Leo sat down again with a thud, struggling to take in the uncharacteristic behavior.

"What is it, Kevin?" Leonardo's brain was still desperately trying to connect the dots.

"Do you want to tell me why I saw a photo and a sketch of Bruce on the news network a few minutes ago?"

Leonardo stood bolt upright, and his hands went up onto his head. "What?! What for? Oh, this is bad," was all that Leo could get out as he started to pace back and forth. "Oh, no, no, no, no, no…" Leo whispered as he continued to walk back and forth.

"Stan, you have to be forthright with me. Now. How the hell did I get wrapped up with a wanted serial killer? And why did you not tell me that Bruce is wanted? He's wanted by the FBI; don't you work for the FBI?" Kevin was breathing heavily as he started to panic.

"Just let me think. Please. Just breathe, Kevin…" Leo held his hand up to Kevin as he tried to work through the information with which he had just slammed him over the head.

Leo didn't stop pacing or look up when he spoke next, "Kevin, you said there was a picture of Bruce? Along with the drawing?"

"Yes, they said that the picture was from the observatory on the day that my two colleagues were killed; they said a witness had taken it before the explosion."

"Oh no, no, no, no, this is so bad…" Leo had his hands on the top of his head again as he breathed and paced, paced, and breathed.

"Look, Stan, I gotta be honest with you, none of this is worth this telescope time in Hawaii. Bruce is supposed to meet us here. If he is wanted, I have to call into the FBI." Kevin was in shock; he felt like he was watching himself speaking, and he felt dizzy.

"No, Kevin, you have to trust me on this," Leo felt calmer as he reminded himself of the purpose of everything they were doing. He pulled his chair in front of Kevin's, leaned in, and spoke in a quieter tone, "The FBI often uses tactics like the one you just saw on the news. Sometimes, they release partial information from an investigation to flush out information from the public. I can guarantee you that Bruce did not kill anyone. This tactic is also used sometimes to give their *actual* suspect a false sense of security, in that the FBI has pinned their crimes on someone else. I know for a fact that Bruce is innocent. I promise you that."

Leo was still looking at the top of Kevin's head; Kevin had not taken his head out of his hands or indicated that he was even listening.

"Kevin, buddy, you've got to stay with me here." Leo sat patiently, still leaning forward. He could see Kevin's shoulders rising and falling faster than usual, and sweat was beading on his forehead around his hands. "Kevin, remember what I told you when we were coming here. *You are in danger.* Can you think about a normal person's reaction after seeing an acquaintance's picture on TV? They would call the police. Think about it. If you call, then they know where you are. If you turn Bruce in, not only are you in danger, but Bruce is in danger."

Leo stopped to let those facts sink in. *It was not even a lie.* Leo continued, "Whoever is pulling the strings by putting Bruce's picture on the news wants you to call, and they want you to turn Bruce in. Don't you see? There are much larger forces at play here than you realize, Kevin."

Leonardo got up from his chair, went into the kitchen, and poured water into the kettle. He turned on the stove element and placed the kettle on it. Reaching into the mostly barren cupboard above the stove, Leo grabbed the box of Earl Grey tea and put it on the counter with two mugs.

"Kevin, buddy, you gotta take some deep breaths. I don't want to have to pick you up off the floor. I'm making you some tea so we can talk this over."

Leo glanced at Kevin and saw him take his face off his hands and sit upright briefly before slumping to the side. Leo rushed over and caught Kevin's upper body as the chair balance gave way, and the chair turned right over with Kevin in it. His head crashed into Leo's lap with a thud on his precariously bent knee.

Almost immediately after falling, Kevin began to

come to.

"You're okay, you're okay," Leo repeated, staying in the precarious sitting position with Kevin's head in his lap. Leo could see Kevin breathing and become aware of his surroundings again. Kevin opened his eyes slowly and realized he was looking at one of Leonardo's feet outstretched in front of his head. He let out a "Gah!" as he tried to sit up, but his bottom was still in the chair, which was now lying on the floor.

"It's okay, it's okay, Kevin. I got you," Leo said. "Let me get out from under you, and I'll help you move the chair so you don't hurt yourself. Just sit up on the floor when I move the chair. We don't want you wobbling all over the apartment, sir," Leo laughed.

Kevin smiled as Leonardo awkwardly shuffled backward to get up from under Kevin's torso. Once he had moved, Kevin fell off balance and ended up on his back on the floor; the chair had flipped again in the move. Both men laughed as Leo pulled the chair back from Kevin's bottom and stood it up.

"There, now sit still with your knees up, and if you

need to, put your head down on your knees." Leo scurried back to Kevin's side and sat on the floor next to him. He put his hand on Kevin's shoulder. "It's okay. Please say something."

Kevin stopped staring at the front and turned to see Leonardo sitting there waiting for Kevin to speak. Leo was ever so slightly leaning in in anticipation of Kevin saying something. *Say something, anything,* Leo thought.

The two men looked at each other for a moment and were smiling at the comical series of events before Kevin leaned in and kissed Leonardo on the lips.

Kevin returned to sitting, and Leonardo stared at him, speechless.

Kevin turned his head away, "I'm sorry, I don't know what that was about."

"Leonardo was stammering, "Don't...don't apologize, Kevin, honestly." He reached his hand from Kevin's shoulder, grabbed him around his chest, and pulled him in for a second kiss.

The two men were motionless for a moment after the second kiss as they stared at each other.

"There," Leo smiled, "We both did it."

Kevin returned the smile, "I was not expecting this to happen, Stan. Please tell me you are who you say you are. I am so unlucky with romance; at least tell me you have been telling me the truth."

Leo's smile vanished. Kevin leaned away from Leonardo and started to stand up.

"Tell me you've been telling me the truth!" Kevin said again, this time more forcefully.

"Kevin, the last several days with you have been some of the most exciting of my life, truly," Leonardo said.

"But..." Kevin winced.

Leonardo chose his words carefully, "But...you know I told you I can't tell you everything. It is crucial for me to keep you safe. But, if I'm being honest...my first name isn't Stan, it's Leonardo. But this is the truth; I have a job to keep you safe."

"Leonardo!" Kevin yelled as he turned away, now on his feet. The kettle was whistling from the stove, and Kevin walked over to take it off the stove.

"Here, let me make you some tea," Leonardo said as he rushed to the stove. "I don't want you to fall again."

"Leonardo. Leonardo…. Leonardo," Kevin repeated the name as if trying to make it sound right. His hands shook as he tried to pull the tea bag out of the Earl Grey tea box.

Leonardo reached out and held Kevin's shaking hand, which now contained the tea bag. He looked at Kevin with his famously intense Leonardo Hoffman stare before he spoke.

"Look at me, Kevin." Kevin stopped and looked at Leo. "I promise you; my name is Leonardo. I promise you; I have a duty to keep you safe at all costs." He squeezed Kevin's hand. "I promise you, if you give me more time, I will tell you more when it is safe. And I promise you, Bruce Hayden has done nothing wrong."

Leonardo let Kevin's hand slide out of his, and Kevin turned and set his now less shaky hands to make tea.

Leo stood silently for over a minute before Kevin finally responded, "I believe you, Leonardo. It's only partially the truth, but…" He stopped and smiled, "millions wouldn't, but I believe you."

Chapter 54: 8-1-8-2

Bruce and AI had devised several workable plans to destroy the carbon copies of his bus ticket before Bruce put AI into power-saving mode. AI used 20% more energy than Bruce would have liked, but he had to be extremely careful while cleaning up any trace of his name before leaving on the bus that evening. AI calculated that functioning in power-saving mode would extend its battery life, including traveling to Hawaii and hopefully getting to Leonardo and Kevin.

The power-saving mode that Bruce had selected gave him the minimum amount of protection via the threat scanner and would only alert Bruce via his watch should there be an imminent threat.

Bruce and AI brainstormed a way to attempt charging AI's battery; however, the resultant crude charger would take Bruce days to assemble. For now, AI was instructed not to use any energy that wasn't one of its critical functions. If Bruce were delayed in getting to Hawaii, he would have to consider building a temporary charging unit.

B.E. Smith

During the last two hours that Bruce was sitting cross-legged beside the dye-soaked fire pit in the abandoned, overgrown lot, he was going over each proposed plan that AI had come up with. He then decided which idea would be primary and which would be secondary to ensure success before catching the bus that evening.

His mental exercise included visualizing his arrival at LAX: checking in, going through security, finding his gate, and boarding his flight to Hawaii. Bruce finished his visualization with breathing exercises, calming his mind and relaxing his body. He slowly opened his eyes and took a deep breath; he was focused.

Bruce packed his belongings and left the remaining ripped-up hair dye box in the fire pit's puddle of hair dye chemicals. He had spent a few minutes earlier in the afternoon burying one of the two bottles he had used to dye his hair. He stood up and put his backpack on, adjusting his ballcap over his newly dyed brown hair.

Bruce walked out of the abandoned lot and into the more open evening air of Chiriaco Summit. He stretched his arms over his head as he got back onto the main street, and he glanced at himself in the first roadside shop window he came to. He barely recognized himself; the brown hair and the tactically placed fist-sized red patch on his face completely changed his look. The red patch on his right cheek had been AI's idea; a simple, irregularly shaped patch painfully etched in his skin's surface layers mimicked an eczema skin rash. It had a dark red color and some peeling skin left around the borders to make it look like an allergic rash, a distinctive marker that people would remember.

Part of the process of branding Bruce's face involved AI interfacing with the nanobots that were circulating in Bruce's bloodstream. Without this communication with the minuscule healing robots to allow the damaged facial skin to stay unhealed, Bruce's red patch etched on his face would have mostly healed within an hour. The exception to the rule allowed Bruce to gain some anonymity for the rest of his trip to LAX and Hawaii.

Bruce carried on down the main street, his ball cap worn low on his forehead like a disgruntled teenager. He watched the swelled number of people walking, talking, and generally going about their days; this would work in

Bruce's favor in his quest for anonymity. He was absorbed in his plan as he walked the rest of the distance to Pete's Gas, which was thirty minutes before Bruce was to catch the highway bus; he lightly tapped his right pants pocket, mentally checking its contents as the gas station came into view.

As he got within fifteen meters of the gas station entrance, he looked at his watch briefly before double-tapping the face of the watch with his right index finger.

Bruce didn't break his stride as he approached the gas pumps and hollered, "Hey there! Nice day!" in his newly minted Southern accent. The large, barrel-chested man who was walking out to the pumps nodded to Bruce and gave a mumbled "G-day" as a large car pulled up into the full-serve lane of the gas station, and the man opened the gas tank with the gas pump already in his hands. This was obviously Pete; his name was neatly stitched on his blue shirt, neatly tucked into his grey work pants.

Bruce could not have timed things better; as the man started pumping gas into the car, there was a loud popping sound followed by an electrical arcing sound coming from the large garbage can-sized transformer sitting on the top of an electrical pole about thirty feet from the gas pumps.

Bruce stopped, turned, and watched Pete as his eyes darted around, looking for the source of the unsettling electrical noises.

"Pop, zap, zzzzzz, poof" were the successive noises from the transformer before it burst into flames.

"Dammit!" yelled the man as he watched the blue and red flames on the top of the transformer get higher and higher. He pulled the gas nozzle out of the car's tank, put it back on the side of the pump, and yelled at the driver, "Sorry! That pole is on fire! Get going so your car doesn't get damaged! Don't worry about the gas right now, Darryl; just go!"

The car roared into life, and the tires squawked as the driver panicked and accelerated out of the gas pump area while nervously watching the transformer light show.

Pete was still looking up at the pole as the car left the station, then he turned toward the building and yelled, "Gail!! Call the fire department! Gail!!"

Gail came running out of the building and skidded to a stop as she took in the six-foot flames atop the hydro transformer before turning and running back into the building.

Bruce moved as he saw Gail running towards the counter he needed access to. He was hot on her heels as she picked up the phone behind the counter, dialed, and started yelling to mobilize the fire department.

Bruce stood off the side of the doors as Gail finished yelling into the phone and then ran back outside to find Pete. She did not notice Bruce standing there entirely in the open.

A convenient side effect of acute stress is tunnel vision. Bruce thought as he made his way to the rear of the counter by the cash register. Once behind the counter, he expertly pulled the mixed bottle of solution out of his right pants pocket, and as he walked past the cash register, he put the nozzle of the bottle in the slit of the bottom drawer of the cash register and squeezed the entire contents of the bottle into the slit.

Glancing outside, Bruce saw Pete waving his arms at Gail. They helplessly watched the fire as melted rubber started to drip down onto the gas station parking area.

B.E. Smith

Bruce found the ticket book copy under the counter, and as he crouched down and pulled the book out, he flipped past the top two tickets that had been sold after Bruce's ticket, and as he saw his last name on the third ticket, he ripped the copy out. Placing the ticket book back under the counter where he found it, Bruce walked back toward the cash register. He glanced around again before entering the code to open the register. Bruce could still hear people outside talking excitedly as more and more people stopped to watch the goings on.

8-1-8-2, Bruce thought as he pushed the numbers with his t-shirt over his finger before pressing the 'open' button at the bottom. The register drawer opened with a percussive "ding!" and sprung open with a "slosh" as Bruce lifted the cash drawer and saw the bunch of paper tickets in the bottom of the drawer covered in water.

Bruce immediately saw 'Willcox, AZ' on one of the tickets and grabbed it, stuffing it into his pocket before closing the drawer again and shuffling back around to the legal side of the counter. Once he was clear of the counter, he glanced around before taking the wet ticket copy out of his pocket, and upon seeing his name at the top, he stuffed it back into his pocket and walked out of the store. The mixed hair solution was quite acidic and would dissolve any ticket copies it came into contact with at the bottom of the cash register drawer.

Thankfully, Bruce didn't have to rely on the chance that the liquid would dissolve his ticket, and thanks to the well-timed distraction, he had time to open the register and grab his ticket.

Outside, Bruce melted into the small crowd that had stopped to watch from the other side of the gas station parking lot.

He could hear a siren in the distance, and moments later, a firetruck pumper pulled up close to the pole, and several men jumped out of the truck and began setting up their hoses. It wasn't long before there was a continuous water stream hissing as the red-hot metal of the transformer met with water. The firemen were yelling at each other as they performed their tasks, and without a break in the water stream, they had connected to a fire hydrant on the following property and had two hoses hitting the transformer from both sides.

A police car pulled up next to where the fireman with the white hat was standing beside the truck, and Bruce could hear the policeman yell from the driver's seat as the white hat went over and leaned into the open passenger window. They talked loudly over the firetruck pump as the diesel fumes and chemical-laced steam encircled the parking lot.

"What you got?" the policeman yelled.

"Oh, it's just a transformer that got too hot," replied the fireman.

"Nothing suspicious that I need to look at?" asked the officer.

"Nope, these transformers sometimes go up in flames when they get too hot over time. Pete says that there was nobody around when it started sparking," the fireman yelled.

The police officer didn't waste any time leaving, and Bruce was relieved to see the police cruiser go down the street and disappear.

Bruce looked at his watch; it was 7:50 p.m. He bounced his backpack back up higher on his shoulders as he waited for his savior in bus armor to get him one step closer to reuniting with Leonardo and Kevin. Bruce looked on as the firefighters ran back and forth, straightening this hose and moving that hose. He was aghast that none of the firefighters were even wearing breathing apparatuses.

Such was the 1980s; firefighters in rural areas like Chiriaco Summit inhaled toxic fumes from all kinds of structure fires and chemical fires year after year, and it was not until at least a decade later that they were required to wear masks at all fires; not just structure fires that they were going into the burning building. The destruction of the human body from inhaling toxic fumes was not understood well, and thousands of firefighters would die of emphysema and cancers before the mask mandate was made as a health and safety precaution for firefighters.

Bruce cringed at the sight of the transformer. Although PCB-filled electrical transformers were banned in 1979 by the Toxic Substances Control Act, most existing transformers in the 1980s were still filled with toxic, cancer-causing PCBs. Bruce and AI had designed the resultant fire from the nano-weapon to be above the compartment of the transformer that contained PCBs so that there would not be a toxic chemical component to the fire. The transformer would have to be removed and replaced when the fire department was finished putting out the fire.

7:55 rolled around, and like clockwork, the bus screeched to a stop well before the gas station as the driver saw all the lights and commotion around the still-smoking telephone pole. Bruce bounded past the fire truck and onto the street as he ran for the bus. The driver saw him coming and opened the door. It was still hissing as it opened when Bruce stepped up the steps with his ticket in hand.

"G-day!" Bruce said as he gave his ticket to the driver. The driver barely glanced at the ticket as he took it and smiled at Bruce as he put the paper under the clip mounted to the side of the driver's area.

"San Bernardino is the end of the line, sir. It'll be around three hours if we don't get held up here," the bus driver yelled back to Bruce as Bruce walked halfway down the seats and sat down, putting his backpack in the aisle seat and sitting in the window seat.

"Much appreesh-ee-ated!" Bruce yelled back in his

fake Southern accent. Only three other people were on the bus; one was at the back, and the other two were between Bruce and the front of the bus. Bruce glanced at his watch and saw AI's scan of the other passengers from the peak of his ball cap.

"No weapons." Bruce exhaled, his shoulders lowered as he got comfortable in his seat. Riding in a bus was so much more enjoyable than walking or hitchhiking.

A few minutes went by, the driver closed the front door, and it hissed to a close. The bus was turning back onto the roadway, and after giving the fire truck a wide berth, the bus made its way down the main street before accelerating as the town of Ciriaco Summit soon faded behind them.

I hope to never see you again, Bruce thought, as his sights were set on the bus transfer in San Bernardino. Then to LAX.

Chapter 55: The Wrath of Shadow

Shadow sat cross-legged on the front steps of the Phoenix Public Library, mulling over the events of that morning. Shadow was frustrated.

I can't believe how archaic these people are, Shadow thought. *I could have outsmarted any of these investigators by age ten. They're like little boys pretending to be detectives.*

Shadow leaned against a large cement pillar to the side of the staircase, looked around, took a small notebook out, and flipped it open, revealing scads of detailed notes. Shadow slowly reviewed all the information, flipping the pages; the hours and hours of video surveillance being replayed in Shadow's head:

"As if we are doing all the work for this case, and the FBI is just swooping in and taking the info they want, whenever they want it," the detective moaned as he sat at his desk; the camera view from the arm of his glasses was broadcasting the crystal-clear image of the detective's desk, handwritten notes, and the seemingly endless line of unhealthy snacks in between him and the other detective's desk nearby.

"Dave, what does it matter, really? We'll do our thing and see if we can catch this fella," the second detective replied, flicking a paperclip back and forth across his desk.

"Marty, it matters because we are doing all the leg work, and the FBI is just riding our coattails. And I completely disagree that they released that photo of Hayden; now he'll go into hiding for sure."

"Yeah, I know. You got all those charges sworn and filed for Hayden?"

"Yup, I coulda charged Bruce Hayden with anything I wanted; that judge barely read the information when he signed it," Dave laughed and shook his head. "It was scary easy to pin that double homicide and car fire on that kid." Dave stopped and flicked a paperclip back onto Marty's desk when it crossed over and spun around in circles before him.

Dave continued, "The Sarge leaned on me hard to

pin all that on Hayden, even though I'm not sure he did all that. The Sarge doesn't like being made a fool, especially with the other carjacking thing. Did you hear that the Sarge was so impressed with this kid that he offered him a job?" He laughed nervously.

Dave looked at the door sheepishly to ensure no one was within earshot. He leaned in and spoke quieter, "Trust me, Marty, I've never known Sergeant James to get egg on his face like that and just let it go." Dave trailed off as another paperclip sailed onto his desk.

"Hey! Keep your stuff on your own desk!" Dave flicked the paperclip off his desk, and it sailed past Marty's desk and onto the floor.

"What about that one witness? What's the deal?" Marty asked as he abandoned the paperclips for a sip of morning coffee. The coffee was still hot; the steam was rising up and around Marty's sizable red nose and soared out into the room as he blew on the coffee before taking a sip.

"I dunno. Comes across as a reliable witness, but there's something about it that feels weird. Too polished.

Cold. Did you get that feeling?" he looked across at Marty.

"Nah, I didn't get that vibe, but I didn't spend any time in the interview room like you did," Marty responded between loud sips of coffee.

"There's something about this that's off, but I can't put my finger on it." Dave shuffled some papers around on his desk, trying to find his interview notes.

"Don't bother, man," Marty pleaded.

"What?" Dave asked as he pulled out a neatly paperclipped set of pages from his interview.

"You know, man, when you get a feeling about something, and you take us off on a wild goose chase. The FBI is sure we have the right guy; why do you have to question things the other investigators agree on?" Marty rolled his eyes as he took another sip of coffee and braced himself for the oh-so-predictable response.

"If you assume, you're gonna make an 'ass' out of 'u' and 'me,'" Dave smiled as he gave his worn-out catchphrase. He put his investigative notes in a neat pile and stood up to put them in his large leather briefcase. He put the briefcase on the desk with a slam and clicked the two side latches, opening it. He placed the clipped package of paper into his briefcase and closed it up, locking the two clips. He stood the briefcase up and looked at Marty as it wobbled on the unlevel, paper-covered surface.

Marty sarcastically rolled his eyes, "where are we going?"

"We're going to talk to Simmons and Graffe about their double homicide that we tied up with a neat little bow for them when we pinned it on our guy," Dave said. "Get your stuff, let's go."

"Okay, man, but let's not get on a huge detour to satisfy your gigantic brain," Marty frowned at the prospect of going around in circles like a dog chasing its tail.

"My giant brain has carried you for years," Dave responded as he smirked, opened the office door, and went out into the smoky police station hallway; one last paperclip fired past his head and pinged off the wall across the hall.

"Yeah, yeah, funny guy," Marty mumbled sarcastically as the two men went out into the hallway, down several offices, and around the corner before Dave knocked once and opened one of the doors that had frosted glass.

"Aha!" Dave yelled as he flung the door open and walked into the office. Getting no response from the two detectives sitting across from each other, Dave walked in and grabbed the lone extra chair in the office, pulling it up so that he was on the third side of the square the two desks made. One of the detectives was slowly typing with one finger on a typewriter.

Dave cleared his throat, "Gentlemen, I've called this meeting...." He stopped when neither Simmons nor Graffe responded—neither looked up from their paperwork.

"Wait a minute," Dave took an exaggerated couple of sniffs, "it smells like lies and deception in here. Maybe a hint of sex?"

"Haha, it's Dave the brave coming here to heckle us about.... let me think. Oh yes, he's going to heckle us about the Neil and Peters homicides." Simmons had barely looked up from his papers, but there was a large smile on his face.

"Heckle you? Come on, Sim the dim, if you can't stand up to my questioning, what are you gonna do if you go to court? Any defense lawyer will put you to shame."

Dave plopped his briefcase onto the junction of the two desks with a thud.

The pencil holder on Graffe's desk fell over.

"Hey now! Don't you have another case to work on?" Graffe yelled as he picked up the pencil holder and put his three pencils back, standing them up neatly. "We already gave you our two homicide charges so you could jam Hayden along with your two geek murders and the car fire.

Although, now that I think about it, I'd rather you have all that hanging around your neck and you be answering to Sergeant James. Right, Simmy?"

He continued as Simmons nodded wisely, "And those charges aren't going to get to trial. Hayden is going to get the gas chamber for his little murdery spree. No wait! Maybe he'll get the chair!" Graffe threw a crumpled piece of note paper at Dave. Dave swatted at the projectile, and it bounced off Simmon's desk and missed the garbage can by a long shot.

"Dave, honestly, do you have a purpose in coming

in here when we are solving some serious crime?" Simmons asked as he looked around his desk for a particular paper.

"Why, yes, Sim, thanks for asking. I want to see the forensic report from the James Neil and Rose Peters murders. Photos and all." Dave unclipped his briefcase and opened it. "Hello," he said in his best Elmo falsetto voice as he moved his puppet hand out from behind the briefcase lid towards Simmons. "I'm Dave's right-hand man, Handy," he continued in his high voice, "and I need to see aaaall your evidence so Dave can save the day and your jobs! Yaaaaay Dave!" Dave laughed at his own joke.

"Okay, handy," Simmons replied sarcastically, "take your moron Dave down to forensics and ask them for the photos because we haven't seen them yet." He paused, "and you're Dave's left-hand man. I hope you have better observation skills than the big hairy ape you're attached to."

"Okay, okay, point made," Dave laughed, closing his briefcase of comic relief. "I'm not waiting for that report; those forensic guys gotta be swamped. Walk me through your scene. You were both there, weren't you?"

"Graffe got there before me. He was there with

forensics the whole time." Simmons mimicked hitting a baseball over to Graffe, "And you're up, slugger."

"Yeah, thanks," Graffe said as he straightened up in his chair. "You have a notepad to take notes?"

Dave pointed to the side of his head, "It's all up here, big guy; just give me every little detail."

"Okay, Davie, buckle up, 'cause I'm not going to repeat myself, you gotta focus. No more fooling around."

"Yeah, yeah," Dave replied as he sat back in his chair.

"Okay, so we got the call after the road guys responded to this house fire at," Graffe flipped to the front page of a large notepad. "8 Fremont Street—the last residence on the block. The guys started knocking on the door of the next-to-last residence, which is 6 Fremont Street, to get them out of the house in case it caught fire, too. They could see a TV on through the window and the legs of someone lying on the floor. They kick the door, and they walk in and find James Neil and Rose Peters, both deceased in the house.

At first glance, it looks like a murder-suicide. The

guys taped off the house and the yard and got the fire department to set up a hose between the houses so we didn't lose the crime scene. I got the call; I hadn't even gone to bed yet that night. I stayed late working on the Stanley murder that I gotta prep for court. So, I go to the scene and take a cursory look while I'm waiting for the forensics guys to get there. At first glance, it does look like a murder-suicide because Rose Peters is dead from a single gunshot to the head, and Neil is sitting on the floor nearby, with a single gunshot to the head.

Neil has a six-shot revolver still in his hand with two spent casings still in the firearm. So, I stood there and just looked around for a few minutes because something seemed off about it."

"Here we go," Simmons groaned, "Detective of the year, Graffe the giraffe, coming through."

Graffe continued without missing a beat. "Anyways, I noticed that Neil had blood on his right hand that likely wasn't from Rose Peters, and it wasn't blood spatter; it looked like he had injured his hand before he died, a defensive wound, I'm thinking, but not from Rose Peters. He also had been hit in the face not long before he died.

Anyway, I got the guys to start canvassing the neighborhood about these two, and they talked to the neighbor around the corner who says Neil lives alone around the corner. So, we get into his house and start poking around, and the one neighbor says that Neil has a car, but the neighbor hasn't seen it around for a couple of days or so. The neighbor says that Rose Peters came to borrow it two days before. So, I get the vehicle info. It's a '76 Impala, and I put out a BOLO for this car, and boom, the guys actually find it parked at one of the motels in the area—the KO Motel.

We tow the vehicle in for forensics to process it, hoping to get something to go on, fingerprints or something. As you can guess, they get nothing. The car was utterly wiped down. I got them to wake up the motel clerk and check the motel office to see who was registered and who was on the list. Mr. G. Cooper. George Cooper, a.k.a. Bruce Hayden. The motel guy is a horrible witness, but he did say that there were two men that he saw; one was Hayden, the other a shorter, stockier, balding man who was significantly older. The other guy never came to the office, but the guy remembered him. They paid in cash when they got the room and must have left sometime in the night, leaving the '76 Impala there.

Now, fast forward to your statement from your witness," Graffe exaggeratedly pointed at the copied report on the top of the pile on his desk from Lee Cain, who saw Bruce Hayden and a shorter, stockier, balding man just outside the observatory before the explosion.

Anyways, back to that night, I don't get that car info until I'm at that scene for a couple of hours, and we don't get the corroborating evidence from Lee Cain until a lot later, but like I said, my feeling was that it was a double murder and not a murder-suicide. Hayden is some kind of pro; he must have had significant military training or something because you don't just wake up one day and start orchestrating murders that look like accidents, like the telescope explosion or the Neil and Peters murders. And surprise, surprise, forensics guys told me they found nothing but Neil's prints on that revolver. The rest of it was spotless.

Now, we've given you two murder charges to go with your other two murder charges, and you can take the car theft charge, too, but Sim and I aren't done with Hayden. I'm convinced he's killed before, and if we don't find him, he'll kill again. Serial killers don't just kill one day and stop cold turkey the next. They get off on it. We want to go back to some cold cases and see if Hayden was responsible. Not too far back, mind you, because Hayden is not that old." Graffe leaned back in his chair, obviously pleased with himself.

"So," Dave glanced back at Marty briefly as he was sitting back in his chair; Marty had been standing inside the closed office door, listening to the details from Graffe. "You can reasonably put Hayden in Neil's car, but you can't put Hayden in Rose Peter's house. What's his motive?" Dave asked.

"Oh, yes, we can," smiled Simmons. "I've got some new info for you; it turns out that Hayden was renting the house next door from Peters. You know, the one that burned down? How about this: he was upset with his landlady, possibly over rent, and killed her and her friend? Rose Peters had already borrowed Neil's car, and Hayden steals it after he kills her and Neil. I got that juicy nugget from asking around at the university about Hayden. You gotta admit, Dave, Hayden has the motive and opportunity to burn down Rose Peter's rental house, kill her and her friend, and go on the run."

Simmons was looking pleased with himself.

"Why not burn down Rose Peter's house then?" Dave asked. "If I were Bruce Hayden, that's what I would have done. If what you're saying is true, what's his MO? Burning, or not burning? We would have gotten nothing but charred bones at that scene. Why not burn it?"

Simmons nodded in agreement, "Yeah, Dave, that would have worked better for him, but what if he never intended to kill Neil and Peters? Maybe he panicked after killing both of them and ran. Even better, what if he meant for her house to burn by being next to his?"

"I didn't know the part about Hayden renting the house next door," Dave mumbled.

"Yeah, well, that's what us real detectives do; we chase leads. Good leads. Not some half-baked, out in left field theories. But there's more, Dave," Simmons continued. His stance had softened because he felt a little sorry for Dave, who was getting schooled. "We did some digging at the university, and he was enrolled in the Physics and Astronomy program at the university campus in Willcox. He started his first year there in September. So, what am I going to say, Davie?" Simmons's eyes were big, his eyebrows teetering near the top of his forehead, exaggeratedly waiting for Dave's answer.

"That Hayden also had the opportunity to know about and kill the two astronomy professors at the Steward Observatory," Dave admitted.

"Yes, Dave, very good. And I won't even rub your nose in the witness evidence you got from Lee Cain. That statement puts Hayden at the observatory at the time of the explosion. Can't you see that this whole mess revolves around Hayden? We get him, then we can rest easy. Now, if you'll excuse us, Giraffe and I have work to do." Simmons pulled his papers back in front of him and didn't look up as Dave and Marty silently left the office.

They walked through the hallways and outside into the back parking lot of the police station in silence. Dave stopped as they walked down the sidewalk beside the parking lot, where several police cruisers were parked.

Uniformed officers were coming and going from the parking lot. Dave turned and looked at Marty. "We gotta go pick up Lee Cain. I have to go over this again. Maybe I'll request a polygraph."

"A polygraph? Marty yelled, "Have you gone around the bend, man? Cain is our star witness; what will happen when the defense gets a hold of the fact that we didn't trust our own witness enough that we had to do a polygraph? I think we have more on Hayden than we've had on people before, and we got convictions on most of those cases."

Dave didn't let up. "It's just so convenient that Lee Cain happens to be at the observatory at the time of the explosion and gives a lame reason for not speaking up right away, oh, and happens to snap a perfect photograph of Hayden's face right before the explosion. Doesn't that make the hair on your neck stand up, Marty? When in your career has something that coincidental ever happened?"

Dave paused to watch Marty start to concede. "I gotta meet with Cain again, Marty, and press for more details because something doesn't add up about the story. I need to know why Cain was at the observatory when the explosion happened, and *accidentally* got a clear photo of Hayden's face. I wonder if they know each other, or Cain is giving us the statement to pin the murders on Hayden...or that the FBI planted Cain into our investigation as a witness because they want Hayden to get the gas chamber."

"Come on, man," Marty belly ached, "You've gone too far out into left field."

The two men silently walked into the parking lot, got into their car, and set out to find Lee Cain.

Shadow returned to the present and watched the blur of people coming and going from the library and walking along the street.

I think it's time to go on the offensive and find Hayden, Shadow thought. *Meeting with those two detectives would be fun, but it appears they aren't going to find Bruce on their own, so Lee Cain needs to disappear, and I need to go on the hunt. Besides, if they can't find me, they'll eventually assume that Hayden murdered me. He won't be able to use that name anymore, not while being wanted by the FBI for four murders.*

Shadow slung the black backpack and started walking down the crowded afternoon sidewalks. After already clearing out the small apartment earlier, Shadow was on the move. For Bruce to continue his mission, he needed to find another telescope and possibly another research partner at a different university. Shadow was deep in thought. Planning.

If only I knew the name of the professor that Bruce has been working with, Shadow thought.

Working with piecemeal information didn't bother Shadow on any other mission, and in fact, the thing that Shadow enjoyed the most in life was hunting down targets. AI had postulated that if Bruce were on the run from the FBI, he would undoubtedly change his appearance and may leave the country entirely.

That meant Bruce would seek the closest airport with the most foreign destination options, LAX in Los Angeles. LAX had high passenger traffic, which would also be desirable for a fleeing Bruce Hayden.

Shadow walked for over an hour before finding a mid-sized car in an otherwise empty parking lot between two old commercial buildings. Breaking in and interfacing with the onboard computer was child's play, and within thirty seconds, Shadow was on the way out of Phoenix.

Now that Shadow finally had complete privacy, the AI unit from the bag was placed on the dashboard of the stolen car, and Shadow and AI began formulating a plan to locate Bruce. The data that would help tremendously was the information about any other stolen vehicles in the area of the KO Motel on the night of the murders. Still, neither the nanocam in the constable's area nor the nanocam attached to the detective had picked up any details yet.

Shadow was already three days behind Bruce, but knowing Bruce's mission protocol in the event of an unexpectedly damaging event, such as becoming wanted by the FBI, Shadow hoped that Bruce had found a safe location, possibly in a nearby town or city to do damage control before deciding where to go.

That meant that, conservatively, Shadow was closer to two days behind. The hunt was what Shadow was born to do, and finding two men of such different ages and descriptions traveling together would be much easier than locating one man traveling by himself.

Shadow wasn't entirely sure if Leonardo Hoffman had died in the desert that day until the police detective broadcast the information via the nano camera that Doctor Leonardo was alive and well. It was unlike Shadow to leave a loose end like that, but the disorientation from the unexpected transport to 1981 left Shadow reeling for hours after arriving. At least the HERB device had been recovered before Shadow left the unconscious Hoffman.

Regardless, it felt good to finally go on the attack after investing so much time in the police's inept ability to find criminals. Shadow had been briefed on what police investigative abilities were in the 1980s, but seeing the police fall all over themselves in a divided way, as Shadow had watched for days, was the last straw.

Shadow looked at the AI display as the city center of Phoenix disappeared in the rear-view mirror, and AI projected the straightforward route from Phoenix to Los Angeles airport onto the windshield. It was 644 kilometers of primarily rural countryside across Interstate Highway 10. After studying the map, Shadow decided that Interstate 10 was almost certainly the highway Bruce and Leonardo would have taken for two reasons: it was the most direct route west within one hundred kilometers either North or South of Phoenix, and the route had the least population and the least number of towns to travel through.

Shadow was aware that George Cooper was dropped off at a highway bus stop by the patrol officer who interviewed him regarding the failed carjacking. When he was dropped off, Cooper had told the officer that he was headed home to Roadforks to see his parents. Shadow, knowing this was a lie, was under the assumption that either the direction of travel was a red herring, or the highway bus was a red herring, or both.

Both AI and Shadow believed that the destroyed car that Bruce was being charged criminally for was a distraction technique to facilitate Bruce's escape from the police that night.

From the map, Shadow thought of Bruce's situation that night and decided that the distraction technique was to pull the police containment units away from the North side of the housing development. It was what Shadow would have done in that situation. After passing through the North side of the housing development, Shadow would have continued out of the hot zone and kept moving West under the cover of night.

Transportation is good, but walking or running is okay when you're being tracked, especially if you're in good shape like Bruce, Shadow thought.

Continuing to look at the map projected on the windshield and looking through the map as needed while driving down the highway West from Phoenix, Shadow continued to think out loud,

"So, Bruce wasn't with Leonardo that night; there is no way that Leonardo could have kept up with Bruce on foot that night. So, Bruce must have been on the move to meet up with Leonardo."

As Shadow spoke, AI intuitively adjusted the map to show the area immediately along the hypothetical route moving West from the North side of the housing development where Bruce was last known to be.

Shadow continued, "Bruce was alone when he foiled that carjacker, then he left in the stolen car that morning from Willcox, alone." AI showed the map where the infamous stolen car was stolen from at the university parking lot. "He could not have known about that carjacker, and so he couldn't have dumped Leonardo off somewhere because the police were right there when it happened. Bruce had to know that Leonardo was already heading West at the time; his direction of travel while being pursued shows his intention to go West. There's a low probability that Bruce would have made his escape with a thought to counter surveillance."

AI displayed its calculation based on Bruce's training. Bruce's mission training was heavy on surveillance, information smuggling, and deception, but Bruce would not have had much training in counter surveillance for this mission. Bruce always had the good fortune in the knowledge that nobody else in this time knew what he was up to, and so being tracked by an assassin from his own time would be completely new territory for him. AI displayed the number '0.05,' or 5% certainty that Bruce made his movements that night with the intent to deceive Shadow.

"AI, we are traveling the Interstate 10 route west to Los Angeles.

There are few enough towns along the way; maybe we'll get lucky and find some evidence that Bruce and

Leonardo traveled this way."

"Copy, scanning."

Chapter 56: doctor Mitch Smith's Dilemma

Mitch Smith had been sitting at home, alone, in the dark, for an hour. Luckily for him, his wife Andrea and their two children had already planned to visit her parents in England for a few days. He saw them briefly in the late evening when he stumbled in the door, drunk. Andrea took one look up and down at her husband and closed her eyes in frustration as she turned to remind the kids to grab their travel cases for their visit.

"Rough day at work, Mitch?" Andrea said through her pursed lips as she brushed their daughter Charlotte's hair into a ponytail and zipped up her one-piece pajamas.

"Uh, yeah, I'm sorry, it was a rough day. I'll get some sleep, and we can holo-chat in the morning once you're settled at your parents." Mitch was staring at the ground, trying his utmost to stay balanced. Andrea hated it when Mitch had been out drinking; it was a source of unending conflict in their marriage.

"Don't bother, Mitch," Andrea picked up their son's

bag and slung it over her shoulder, "get your head on right before we get home on Tuesday. Please?"

She turned and gave Mitch a knowing look before kissing him on the cheek, "I know, you can't talk about it."

"Sorry, no, I can't," Mitch replied as he watched Andrea, Charlotte, and little William walk out the front door and turn towards the bus stop at the corner.

The bus would take them to the high-speed train, then transport them to London in around two hours. It was Andrea's preferred time to travel since the kids slept on the train and would get to her parents in the middle of the night. Mitch watched them walk out of sight, and he slowly turned to face his empty apartment. He closed the door and locked it behind him, and he stumbled to the couch and sat down with a flop.

"Ohhhhhhhhh!" Mitch groaned as he sat with his eyes closed. *Okay, Mitch, get yourself together,* Mitch thought. *Okay, I know what I am supposed to do, but how do I know that Leo didn't create this as an elaborate setup of some kind?* Mitch ran his hands clumsily through his sparse curly hair as he thought.

"Tea, I need tea," Mitch said aloud as he looked at Jeremy, their kitchen robot.

"Jeremy, I need some tea. Lemon ginger tea and a Tactol, please. I need to sober up."

Jeremy, the kitchen robot, sprung into life, setting water in the kettle to boil, and was noisily rummaging through the cupboard; apparently, the lemon ginger tea had not been used in a while.

Mitch ran his shaky hands over his head again. *I have to follow through like I promised, and Fran will absolutely murder me if I don't do what I'm supposed to. A different job seems like a good idea, where my actions don't directly impact the whole world. Fran is so scary.*

Mitch shook his head in a weak attempt to shake the alcohol that was impairing his brain.

Okay, think, think, think, think... Mitch was struggling to think.

"Lights off and darken windows." Mitch groaned as he noticed the light was one source of his discomfort. It was hurting his head. The lights turned off, and the filaments in between the panes of glass in the windows became opaque, blocking the outside nighttime street lights. All the light that remained in the house was the soft glow of lights on different electronic devices around him. "Much better," he said.

Jeremy walked over to the couch in the dark and held out the teacup and little blue pill on a small plate. "Your tea and meds to sober up, Mitch," Jeremy said, "ginger-lemon tea and fifty milligrams of Tactol."

"Thanks," Mitch sighed as he took the teacup off the tray in one hand and clumsily grabbed the little blue pill with the other hand. Jeremy dimly lit the tray as Mitch searched for the medication.

"You're welcome," Jeremy responded, and he walked back to his spot in the kitchen and powered down.

Mitch blew over the surface of the hot tea to cool it before he took a sip and downed the blue pill. It would only take about thirty minutes for the Tactol to react with the alcohol in Mitch's body and change the molecular structure of the alcohol so that it no longer impaired the brain. The only issue with the Tactol was that he had to deal with his non-numbed emotions as it sobered him up. The Tactol also dramatically softened the blow of hangover symptoms, a convenient 22nd-century invention.

The mixture of ginger and lemon soothed Mitch as he sipped but reminded him of all the other times he was in this situation: drinking to cope with work.

As Mitch sat in the dark, sipping his tea, he went over the steps Leonardo Hoffman had instructed him to memorize if the mission had been compromised. He put his tea down on the small wooden table beside the couch and sat back in his seat with his eyes closed. The darkness felt nice on Mitch's thumping headache. The fateful day that Leo gave Mitch the instructions he was following played in Mitch's head:

"Do you understand?" Leonardo leaned in on his chair towards Mitch; his eyes showed the sheer piercing intensity that Leonardo was famous for. Mitch squirmed in his seat.

"Yes, I get it," Mitch responded as he wiped a fresh bead of sweat off his forehead, "you want me to memorize these things if the mission that I don't know about is compromised."

"Mitch, look at me," Leonardo was sitting directly in front of Mitch, breaching Mitch's comfort zone.

Mitch looked Leonardo Hoffman in the eyes. He spoke slowly and deliberately.

"Mitch, you are about to start work on a mission that will change your life. This mission will change life on this planet as we know it, regardless of whether we succeed or fail. I have to know that the people I am bringing on to be a part of this mission will trust that I know what I'm doing. This mission is the culmination of my life's work, Mitch."

"I trust you, absolutely, Leo. I can't think of any other person outside of my family that I trust more than you. Your integrity is second to none, and I have no doubt that you will always do the right thing."

Leonardo let the compliment sail by unacknowledged. "I need you to be able to trust me, even if other people outside our small research group say otherwise, even if something were to happen to me and it looks bad." Leo was still staring intently at Mitch.

"Leo, you're freaking me out here," Mitch sat back in his chair to give himself some space from the encroaching scientist. "How could you say that you suspect something will happen to you? Are you expecting something to happen to you? We're scientists, for crying out loud. We're not spies or secret agents."

"Mitch, listen." Leonardo sat back in his chair to ease up on the suffocating Doctor Smith, "this mission is going to be unlike any other you've ever heard of or read papers about; this is beyond huge. That requires that I safeguard the mission from any and all sabotage or interference that I can think of. I have to anticipate any problems we may have and develop contingencies." Leo paused. "This is just one of those many contingencies. Mitch, I hope that this never comes to pass, but we must be prepared if it does. Do you understand?"

"Yeah, I get it, Leo. You're either being uncharacteristically dramatic, or this is unimaginably big." Mitch was sweating profusely. *What the hell could be so big that even Leo says it will change the world?*

Leonardo, anticipating Mitch's confusion, put his hand on Mitch's bouncing knee.

It stopped shaking briefly. "Mitch, everything will

be revealed once we start this mission together. I need your word that you will commit this series of tasks to memory. Swear it." Leonardo removed his hand and sat upright in his chair.

"I swear," Mitch said sternly. "But you better not be pranking me, Leo; you're talking this mission up so much, it better be good, or I'm going to be disappointed."

Leonardo Hoffman sat and grinned a wide, toothy, slightly manic grin. "You won't believe it until you see it, my friend." Leo paused momentarily as he looked at Mitch intently, "Mitch, commit it to memory. Today. If this should come to pass, that phrase will be spoken to you, or Gus, or Fran, in my voice, and no matter what is happening around you at that time, you do this series of tasks. It must all be completed within thirty-six hours of hearing that phrase, not one minute longer."

Leo stood up and patted Mitch on the back, "You can't talk about this again to anyone, even me. Ever." Leo gave Mitch's shoulder one last squeeze, and he opened the door and walked out of Mitch's office, shutting the door behind him.

Mitch then recalled the memory of the tasks he had been given by the brilliant scientist years ago; Mitch did as he was asked and committed the sequence of instructions and events to memory:

Assume that you will be under constant surveillance.

Leave the lab and go straight to Rolli's pub. (Mitch liked that one since alcohol was his go-to when seriously stressed).

Sit at the bar and order from the robot tending the bar; not a person, the robot. Ask the robot for the 'Coach's warm-up brew.' Make sure you pay using the robot's retinal scanner.

After your first drink, go directly to the Oliver Twist Bar across the city. Take the bus. Have six alcoholic beverages of your choosing, but you must first make small talk with the robot; use the phrase, 'My left hand and my right foot are numb,' before you order your first drink.

After your six drinks, go home and have Jeremy make you ginger-lemon tea and get you a Tactol.

You will receive further instructions shortly after you consume the Tactol and all of your tea.

Mitch shook his head and sat bolt upright. "Oh no! I forgot to drink all the tea!" He grabbed the almost full tea cup off the side table and drank all of it. It was lukewarm. His head was spinning with anxiety, and thankfully not alcohol, after the Tactol had rendered the alcohol ineffective in his body.

Mitch clumsily put the empty tea cup back onto the side table and sat at the front of his seat on the couch. He shook his head at the absurdity of what he had done; he followed all of the steps to the letter, and now, was he supposed to have some epiphany? He laughed nervously as he looked around his dark, empty apartment.

What did that crackpot old fool do? Mitch thought as he waited on the couch. Mitch sighed as he looked around, not knowing what to expect from the mad scientist's instructions.

All of a sudden, his stomach began to gurgle, and Mitch leaped up with a shout, "Ah!" and he ran for the bathroom, just making it to the toilet in time to vomit the entire contents of his stomach in two shocking minutes of vomiting.

Afterward, Mitch sat on the bathroom floor, his stomach muscles still reeling from the violent purging of his stomach. He felt absolutely ridiculous to follow Leonardo Hoffman's instructions unthinkingly.

He was reaching for the toilet handle to flush his surprisingly fluorescent-colored vomit when his eyes were drawn to movement in the toilet bowl. He leaned farther over the toilet as he looked in surprise. There, before his eyes, several words were being spelled out in the toilet water.

"Close the door and lock it," the words in the toilet bowl spelled.

Mitch, in disbelief, turned around and locked the bathroom door closed. Immediately, there was a holographic projection in the small bathroom space, and Mitch sat back down on the floor with a thud as he saw the virtual projection of Leonardo Hoffman. The holographic Leonardo smiled for several seconds before speaking:

"Mitch, I see that you have done what I asked; thank

you for that." Mitch just stared with incredulity.

The ghost of Doctor Leonardo Hoffman continued, "Mitch, I'm sorry for all the layers of secrecy, but if you're watching this, the worst has happened: a mission compromise. These nanobots of mine in your toilet are shielding your bathroom from any other electronic surveillance device, and they have confirmed that I am speaking to you, well, from checking the DNA in your stomach. Sorry about that, but I did make the vomiting as pleasant as possible, with the ginger-lemon tea and all."

"You bugger," Mitch mumbled as he watched the mischievous smirk on Leonardo's face. He couldn't help but grin at the genius of it.

"I needed to get those nanobots inside your stomach, Mitch, and I couldn't think of a better way for you than taking them with a few alcoholic drinks, no?" Leo smiled at the truth, although Mitch could tell Leo was about to get serious.

"First of all, I'm sorry that you're in this position to be helping me without me being there. You are a loyal friend, and I value loyalty above all else." Mitch felt a sudden wave of guilt wash over him; the argument in the lab before Leo disappeared. Leo continued, "Mitch, I need you to go into the lab and do something for me. These instructions are to be followed to the letter; memorize them, and do not record them or speak of them outside this bathroom. You must assume that your every action is being monitored around the clock once you leave this electronically shielded room. I need you to go to work for your next shift, as usual, and complete your everyday repetitive habits like making coffee, getting your computer set up for work, having a quick chat with Fran, and heading to the bathroom.

After you have done your regular routine, tell Fran you have to check the lab supply room for some encrypted memory devices to back up your current device. Go to the supply room and let the door close behind you. Leave the light off. Find the cleaning robot in its storage harness; don't activate it; just feel it with your fingers and put your current memory storage device in the robot's front chest port. Count three seconds, remove the device, and put it back in your pocket. Turn the light on, open the bin with the spare storage devices, and take one to back up your current device.

If asked, you dropped your device on the ground when you entered the supply room, and the door swung closed behind you. You picked up your device and then turned the lights on before getting your replacement device.

Then, return to your desk and use your computer to transfer all the data from your old storage device to the new one. Report that the automatic light in the supply room is malfunctioning. Dispose of the old device as per protocol." Leo smiled again, "Clear as mud, my friend?" Like all of these tasks, you are the only one who knows your role, and it needs to stay that way. You have to trust that I know what I'm doing."

Leo's face softened somewhat before he spoke again, "Mitch, I can only imagine the stress you are under and the horrible circumstances that have brought this contingency plan into action. You have to know that I am doing my best." He stopped. "Or I did my best, depending on what happened to me. I hope to see you again, my friend. Good luck." Leonardo's projection vanished, and Mitch was left sitting on the bathroom floor.

Now completely sober and alert, Mitch flushed the nanobot message down the toilet, went to the sink, splashed water over his face, dried off, and left the bathroom. The series of tasks were playing repeatedly in his mind as he changed out of his clothes and got into bed. He looked at his watch. It was almost midnight.

It's been five hours since the message was delivered to Fran on the phone in the lab, so when I go into work Sunday morning, it will be almost thirty-five hours post-message; nothing like giving me enough time to do what I have to do, Mitch thought as he shook his head again at Leonardo's ability to pull strings, even when he wasn't there in person. *I hope you're not dead, and you know what you're doing, Leo.* Mitch laid his head down and closed his eyes to get some sleep; his body was exhausted, but his mind continued to race throughout the night as he tried to regain some sense of control of the situation in which he found himself.

Chapter 57: Making Due

Leonardo woke first on Sunday morning; it was the only day of the week that Kevin had rest from his telescope research time. Sunday was the day to reset and get the food and toiletry supplies for the coming week. Leonardo stretched and wiped the sleep from his eyes as he brushed his teeth in the bathroom. Looking at himself in the mirror, he shook his head at the stupidity of what he had done. He had risked the entire HERB mission, not only by traveling back to the year 1981 in a split-second desperate decision but also by becoming romantically involved with Kevin, who was the most significant part of the mission outside of Leo and Bruce.

He looked at the dark circles under his eyes and the wispy, greying hair on the top of his balding head. He had spent most of his childhood and his entire adult life in the pursuit of scientific discovery, only to be thrust into an active mission at the pinnacle of his scientific career and completely disarmed by his idol, doctor Kevin Nault. No matter how Leonardo examined the circumstances, they were mind-bending in their complexity.

Should he back away from Kevin while Bruce was still missing and risk further alienating Kevin? Or should he continue his path and steadily push his way into Kevin's research to see how he could assist in doing the tasks that Bruce was supposed to do? Leonardo had the fully functional AI unit with him; therefore, all the information Leo needed was accessible to him to help make the asteroid discovery, which was, at best, nine and a half months away. How would he be able to insert the necessary 22nd-century technology into Kevin's research while allowing Kevin to think that he helped create it?

His façade as an FBI agent had taken him entirely out of the equation regarding having any believability as a research partner; his mission had become security, and going from security to research didn't seem plausible. Kevin would know that something was not right.

Kevin had already suspected that something wasn't right about Leonardo, and even though Leo had told him his real name, he did not divulge anything further. Leo knew about the previous mission, during which Bruce had made the mistake of telling Kevin the truth about the HERB mission, and the results were disastrous. Kevin could not know the whole scope of the mission at all costs.

"Good morning, sir." Kevin sang as he passed by the open bathroom door and looked in at Leonardo, staring into the mirror; toothpaste was all over the sides of his mouth like a rabid dog.

"Uh, good morning, sir." Leo talked through his mouth full of toothpaste as he lost his train of self-talking thought. He had been startled by Kevin while he tried to work out his existential crisis via the mirror. Leo and Kevin had taken to calling each other 'sir' as a friendly inside joke name for each other.

Kevin walked into the kitchen area and started the kettle for his morning tea. "Tea?" Kevin looked over at Leonardo as he entered the small living room and looked out the window at the barren landscape outside.

"Please," Leo smiled as he continued to look outside.

"What's up? You seem preoccupied," Kevin said as he continued to rummage around the kitchen to get some breakfast.

"I'm just worried about Bruce. The police have cast

a wide net to get a hold of him. I know he didn't do anything wrong. I wonder where he is." Leo was still looking out the window.

"Why are you so worried about him?" Kevin asked as he had turned to face Leo's but was talking to his back.

"I'm worried because he is my responsibility, and I would like him to get here in one piece. Sooner than later would be preferable." Leo turned and saw that Kevin had stopped what he had been doing and was looking at him.

"Bruce is a super-intelligent young man, and there is something about him that is amazing, but I can't put my finger on it, you know what I mean? I can't fathom how or why he became involved in this whole mess, and now he has to be kept in hiding by the FBI, from the FBI. It's head spinning."

Leo answered swiftly to avoid having another long-drawn-out conversation about why Leo was trying to keep Bruce safe while his own company told the public he was wanted for four murders.

"Bruce *is* an amazing young man; I agree completely with you on that. You have to understand that there are different branches or arms of the FBI that are completely separated from one another. One branch wants Bruce for questioning, and another branch," Leo motioned to himself, "a more covert branch, wants to keep Bruce safe in hiding for now. It is a national security operation."

"Uh huh," Kevin said sarcastically as he returned to find his ingredients in the sparsely stocked cupboard.

"I want to be as upfront with you as I can about this, Kevin," Leo continued, "when Bruce gets here, he will most certainly be using an alias, and he may have altered his appearance on the FBI's advice. It is very, very important that nobody knows that he will be here with us."

Kevin's back went ramrod straight at the stove as an idea struck him. He raced over to his briefcase and rifled through it. Multiple papers fell out as he was searching for something.

"What is it?" Leo asked.

Kevin didn't answer until he held up a piece of paper with an 'aha!'

"My letter to the university to request that Bruce be able to take a leave of absence from his program to help me with my research."

"So, you didn't tell anybody?" Leo asked.

"No, I typed up the letter, and I was going to submit it the day you came to meet me outside my house."

"Did you talk to anybody about it?" Leo asked.

"Nope, nobody." Kevin dropped the folded paper, which floated onto the small coffee table beside him.

"Okay, great." Leo sighed a breath of relief.

"Although, if any police investigator goes to the university, someone may have seen Bruce meeting with me on Saturday mornings; I can't say for sure if anybody would have put it together. All they would get is that Bruce was enrolled in the Physics and Astronomy program, along with his home address, and I would be listed as his professor for a couple of his first-year courses.

Leo shifted the subject again, "Kevin, I was thinking that maybe while Bruce is not here, I could help you with your telescope time. I mean, I'm not Bruce Hayden, but I excelled in Math and Physics in High school. Maybe I can help do some monotonous tasks to help you out."

Kevin smiled momentarily before answering, "Uh, sure, I think I could find some stuff for you to do to help things move more smoothly with my research, but they won't be glamourous, Leo."

"Great, I'll start at the bottom, as it were," Leonardo laughed. Besides, it'll be good for me, too, since I will get to keep an eye on you during your research time and not be sitting here watching TV. Or reading one of those horrible romance novels that someone left behind here."

"I know, right?" Kevin laughed, "Those are so horrible!"

Kevin changed the subject this time, "I'm going to take a cab down the mountain whenever it gets here. I already arranged for them to come at 10:00; I called last night so I didn't forget this morning when I was in a fog. I've got quite a list of groceries that we need for the next week. It's too bad the workers who run the tuck shop are gone for a few weeks. Otherwise, we could get a lot of what we need from them. You want to come with?" Kevin smiled mischievously. "I know, I know, you *have* to come with me."

"Yeah, sorry, I do." Leo responded, "I hope you don't get sick of me before long."

"How could I?" Kevin asked as he motioned to everything around him; Leo thought it was a little sarcastic. His smile faded, "Leo, honestly, let's just enjoy our companionship before Bruce gets here, and we have company for the rest of the year."

"No worries about that, Kevin. I already have thought of that. You keep your room, and I'll move from the second bedroom to the couch, and Bruce can have the second bedroom. I'm sorry, but I can't split the three of us up once Bruce gets here for security reasons."

"I understand," Kevin said as he looked at his watch and started to gather his keys and wallet in anticipation of the cab arriving.

Leo watched as Kevin scurried around, getting his list and keys and checking his wallet's contents as if something had shifted since he used it last. "Kevin, I wanted to tell you that I'm glad that we've met; it might not be the best circumstances, but I never had any intention, nor have I ever become involved with anyone in this way before you. I just don't want you to think that I'm a fly-by-night, romantic FBI agent."

"I know what you're about, Leonardo. I know you're not just interested in me because we spend most of our days together, and there aren't many other options, to be honest," Kevin and Leo laughed.

"Well, doctor Gilbert, who has the cross shift, is putting out a vibe if I may say so," Leo joked. "But you're right; there are no other options up here on the volcano."

As the two men joked back and forth, the blue cab came into view and kicked up the red, Martian-colored sand.

"Right on time," Kevin said as he glanced again at his watch, "I guess they get this call often throughout the year. They've certainly got the timing down pat."

Kevin and Leonardo walked out of the apartment and into the crisp Sunday morning air. Leo breathed the air deeply; it felt nice on his lungs, which were slowly but surely acclimatizing to the thin volcano-top air. Kevin got into the back of the cab, and Leo entered the front passenger seat.

"Ah, Kea-Kao! Nice to see you again!" Leo announced as he got into the front seat and recognized his friend from several days prior.

"Hello, Stan! My friend!" Kea-Kao responded with a big belly laugh.

As the car with the three men descended the volcano, and they twisted and turned, following the contoured roadway, Leo mentally reviewed the information he had gotten from AI early that morning before Kevin got out of bed.

Leaving the tiny scientific community at the volcano top stressed Leonardo. So, in preparation, he had outfitted himself with one of the remaining nano cameras, along with the portable AI unit, which was equipped with a tiny earpiece so that AI could discreetly talk to Leo on their journey for groceries that day.

Leo was supremely uncomfortable after Bruce failed to make the flight to Hawaii, and now, after almost a week, there was still no sign of the young mission specialist. Leo had planted the seed in the AI unit to analyze the grossly deviated mission and determine what Leo's course of action needed to fill in for Bruce and, in the worst-case scenario, continue the mission without Bruce.

He knew that after Bruce had gotten him the napkin message on the plane, communication would be difficult, especially now that the FBI had released Bruce's photo to the public to locate him. AI had confirmed for Leonardo that the image of Bruce released by the FBI had to have come from one of the nano cameras that were positioned outside the Steward observatory leading up to the destruction of the telescope and the murder of two scientists. That meant that Shadow had somehow given the photo to the police, which implicated Bruce in the two murders; it was a bold move, but in terms of a disruption technique by the 22nd Century operative, it was a good move.

Shadow's disruption of the mission to date meant that the name Bruce Hayden and the Steward telescope had been rendered useless. Leo had also asked AI to compare Shadow's intervention locations on a timeline with the level of information to which Shadow must have been privy.

In other words, Leo wanted AI to postulate how much information Shadow had before piggybacking Leo back to 1981. Shadow knew the importance of the Steward Observatory, but, for some reason, he felt that destroying Bruce's house by fire and murdering Miss Rosie and Mr. Neil was a better tactic than attacking the second most important person in the mission: Kevin Nault.

AI calculated that it was a certainty that Shadow knew Bruce's Willcox address before arriving in 1981. The layers and layers of security and redaction during previous mission debriefs meant that only a tiny circle of people knew locations, names, and dates from the mission. The AI unit used in the earlier missions was treated with the highest security possible. Still, some information about Bruce's residence and the Stewart telescope had been leaked to Shadow. Even in debriefing prior missions, Leo could only think of a couple of people who may have heard Kevin Nault's name besides Bruce.

When Bruce came back from his two previous unsuccessful missions, as soon as he arrived, his AI unit was encrypted before the sensitive information was copied for extensive analysis. The encryption allowed for dummy names to be used in the place of people and places. For example, the name Kevin Nault had been replaced with Joey Hart for all of Bruce's debriefings with psychologists and behaviorists and for official reports.

The only people who were *supposed* to know the name Kevin Nault were Bruce and Leonardo, and ever since Leo recalled the faraway meeting with the board of directors that day on the plane to Hawaii, he was suspicious of the director for insisting that he had access to all of the highly sensitive information.

In Leo's mind, there was nothing to gain by allowing the director to know real names or real places in the mission; however, the director had taken it to a vote that day and had been shut down in a second vote after the first vote had ended in a deadlock.

There had been a significant leak in the sharing of highly encrypted mission information, and the presence of Shadow in 1981 proved that. Shadow seemingly had some sensitive information, but not all information, which was perplexing to Leo, and his mind was working around the clock to figure out why. AI had been computing tens of thousands of scenarios and hypotheses that morning and would have a report for Leonardo when he could arrange private time and inspect the data without fear of being caught.

"You good Stan?" Kevin asked as he tapped Leo on the shoulder from the back seat.

"Oh! Yes, I'm good.... you?" Leo stammered as he returned to the present, and the cab wound around the last couple of turns before getting down to sea level.

"Never better," Kevin smiled at Leo; the sarcasm was palpable as the two men exchanged knowing looks at their precarious situation. Kevin, oblivious to the complexity of the problem, felt he was being charming, and Leo smiled back.

"Nanocams at sea level are clear," AI quietly chimed deep into Leonardo's ear canal with its female human synthetic voice. The voice of the supercomputing device was indistinguishable from a human, and Leo had changed AI's voice to female from male that morning; he felt more at ease listening to crucial information from a female than a male.

He wasn't sure why, but he had realized soon after

getting to Hawaii with Kevin that he liked his AI to have the softer voice of a female. Leo smiled at his thought process, and as he glanced back at Kevin, whose hand had lingered on Leonardo's shoulder, Leo thought for a moment that he wanted AI to speak to him in a female voice so that another male voice wouldn't distract him from hearing Kevin's voice.

Leonardo had a few long-term relationships with men over the years, but science always was his passion, and relationships came and went like waves lapping up onto the beach of science and then receding into the ocean. Work always won over Leo's relationships. Always. He had been singularly focused on studies and research from a young age, and no distraction had ever threatened to break that focus.

"Here we are, sirs," Kea-Kao sang as he turned onto the bustling street that housed the small grocery store.

Leonardo shook his head in an attempt to stop daydreaming, and of course, Kevin noticed. The car pulled up a few car lengths from the front of the grocery store, and Kevin handed the cab driver the money for the drive.

"Keep the change, Keah-Kaoh," Kevin said as he half-butchered his name, "Let's go dreamy head," he continued as he got out of the rear of the cab first and gave a slight bow as Leonardo sheepishly got out of the cab.

"Kea-Kao, don't go!" Leonardo yelled as Kea-Kao pulled the steering wheel gear shift into drive.

"You wait right here; we won't be more than fifteen minutes, and you can bring us right back up," Leo directed as he closed the front passenger door. He leaned in to make eye contact with the driver to confirm that the driver would wait, then he turned back towards Kevin with a stressed smile.

"Right this way, sir," Kevin continued with his grandiose body language.

"Okay, okay, funny guy, let's get our stuff and get out of here. We can't stand out to any of these people." *Don't forget who is hunting us*, Leo thought.

"You're such a buzz-kill," Kevin retorted as the two men walked through into the front of the open-air market.

AI was scanning continually and would alert Leo of anything suspicious. Leo had spent hours and hours preparing with AI for this visit to the town for supplies, but it didn't make him any less nervous. He watched everyone coming and going as he and Kevin wound their way around the small marketplace. Leonardo concluded that he was sweating more than anyone else around him, and as he noticed his super sweating, he began to panic and felt like he was sweating even more.

He looked at the ground, took a supersized breath, and exhaled slowly to calm his overactive nerves. He had one paper bag under his right arm that Kevin had handed him as Kevin expertly visited all the little booth-type merchants within the covered market space. He could smell the fresh plantains and a sweet-smelling fruit that wafted up from the bag to his nose as he followed Kevin around.

Suddenly, there was a loud metallic crash from about fifteen feet away to Leonardo's left. He jumped and turned towards the noise with a loud "AH!" which was immediately followed by muffled laughter from the people who watched him jump straight up in the air as the metal tray being washed at the cured meats booth fell to the ground. He was the most anxious person in the market.

He looked past an elderly couple smiling at him, and

the man who dropped the tray gave the universal thumbs-up signal, more as a question. Leo nodded at the man as he jostled his paper bag of groceries and checked the ground to see if he had dropped any. As he looked back up, he caught Kevin a short distance away, laughing so hard that their faces matched in color, and tears ran down Kevin's face.

As the two men got within speaking distance, Kevin leaned into Leo and said, "Well, that was memorable, don't you think?" Leonardo busted up laughing as the two men continued through the busy market, eventually escaping all the eyes that Leo had attracted with his anxious outburst.

They got to their penultimate vendor, who sold Kevin several spiral-bound notebooks, a box of HB pencils, and a yellow highlighter. As they left the open-air market, Kevin gave Leo the bag he was carrying and pointed to a store two doors down from the parked and waiting cabbie.

"I'm just running in there for some dry-erase markers; sadly, there is no chalkboard here, so dry-erase will have to do." Kevin veered off at a trot as Leo looked this way and that, unsure if he should follow or get to the cab.

Not wanting another embarrassing reminder of his terrible security skills, he slowly walked up the skinny sidewalk to the front of the small shop. After seeing Kevin happily chatting up the young lady behind the counter, he turned and started to watch for an opening in traffic to cross over to their waiting ride. Leo got to the cab and waved at Kea-Kao to open the trunk, and after neatly placing the several bags of food and supplies, he closed the trunk and got into the back of the cab.

This will give me a little more privacy on the way home, and I can see Kevin at all times, Leo thought.

After exchanging pleasantries with the driver, Leo looked down at the watch on his wrist and tapped the code pattern onto his watch for AI to give him the current scanning properties and list any anomalies located. Leo knew AI would immediately notify him of any imminent threats, but checking in this way allowed Leo to know the full scanning results as they wove their way through the market and back out to the street.

"Scan from 11:08 hours until 11:25 hours this morning: many sharp and blunt metal objects used to prepare meats and fish were detected at eight locations within the marketplace. Fourteen people, nine males, and five females, carried small fold-up knives in pockets or on belts. One firearm was detected at 11:18 hours being worn in a holster by a male with a metal 7 cm wide and 8 cm long badge-shaped object in his pocket; deduction: a police officer or armed security person. At 11:19 hours, your heart rate spiked to 160 beats per minute for 17 seconds, and your blood pressure peaked at 175 millimeters."

Cripes! Leo thought as he tapped his watch once to stop the embarrassing statistics of his poor heart health and his on-edge startle response. He felt his face flushing again. *How the hell am I going to pull this off?* he thought. *Bruce, you need to get here ASAP because I'm drowning. I'm playing an incompetent FBI agent. I'm not going to be believable if I try to give input into Kevin's research. The asteroid discovery is only a few months away, and I'm falling for the only guy who can make that asteroid discovery believable*. Leo had his face in his hands when the front passenger door of the cab opened, and Kevin got in clutching his four dry-erase markers.

"Ready?" Kea-Kao asked as he put the car into drive.

"Yes!" both Kevin and Leo exclaimed at the same time.

"So, when did I get promoted to shotgun?" Kevin smiled as he turned his head and gave Leo a side-eye glance.

"Shotgun?" Kea-Kao asked as he looked nervously at Kevin.

"No, No, not the gun; it's an expression," Kevin answered with his hands out to try to ease the driver's nervous look. "The expression' riding shotgun' is actually from a book by Alfred Henry Lewis, written in 1905, and refers to the front passenger having a shotgun while riding the stagecoach that was full of money or goods." Kevin paused a moment.

"And that book was called?" Kevin asked as he gleefully looked out the front of the car. They were within minutes of beginning their ascent back up the volcano to their research facility.

Leo started smiling like a fiend as AI unexpectedly joined the game and announced the name of the 1905 book into his ear.

"The Sunset Trail," Leo chanted with a big grin, repeating the words that AI had just given him.

Kevin looked back at Leo with an impressed look on his face.

"Impressive, sir. A professional *and* a well-read individual. Let me give you a follow-up question: What pseudonym did Alfred Lewis write under?"

Leonardo smiled as he waited for AI to play along, and his smile became more prominent as AI was stumped.

"Unknown," was the response given in Leo's ear.

"You've got me on this one, Kevin. I don't know." Leo responded, still smiling at the interesting situation he found himself in.

"He wrote under the pseudonym Dan Quin," Kevin proclaimed. "But I'm impressed that you knew what the book was called."

AI again joined the conversation, surprisingly in Leonardo's ear. It was like having a buzzing bee flying in his ear as he tried not to react to AI's launch into Dan Quin's other works.

"If you are interested in Sunset Trail, Alfred Henry Louis also wrote Confessions of a Detective, in 1906, and..."

Leonardo shook his head in discomfort, as AI had completely hijacked his ear. Kevin looked at Leo curiously, tilting his head. Leo stopped shaking his head. Kevin turned back towards the front of the car with an incredulous look.

"Er, I think a bug flew in my ear! How annoying!" Leo lied. He tried to join the human conversation again, "well, I like trivia, but I can't say that I've actually read that book," Leo admitted as he felt pangs of guilt rising for cheating at the game. He made a mental note never to allow AI into further conversations with Kevin and not let AI bait him into lying to Kevin. There were enough lies to keep straight, and Leo was already a horrible liar.

Leo spent the rest of the trip back up the mountain formulating some semblance of a plan to re-introduce Bruce into the Leo and Kevin show. He knew there would be hard questions from Kevin when Bruce did arrive, and although he had done pretty well so far, he was looking forward to seeing Bruce's face again. He would have to muffle his excitement when Bruce arrived, lest it elicit more questions from Kevin.

Please be alive, Bruce, and please get here soon, Leo thought.

Chapter 58: Transportation of People, Goods, and Energy in the 22nd Century

The 150 years adjoining the 1980s to the 2130s in America were not smooth by any stretch; however, after America's rise from the ashes of the Great Civil War in the 2060s, America became unitedly focused on what was needed to propel her into the 22nd century. Commercial transportation of goods had long been tested and perfected in Europe and Asia before America built her new infrastructure from the ground up.

High-speed magnetic suspension electric train lines between major cities powered by massive solar electric arrays were built first. Offshoots of those significant lines were created to service the population not in major city centers. The resultant transportation map of America was vastly different, with trains moving most people.

Long, enclosed tube-like sections of train tracks allowed the trains to travel at speeds up to 600 km/h, and ingenious train 'passing areas' were incorporated from the start to allow multiple trains in opposing directions to use the same line simultaneously. Automated train traffic controllers allowed the trains to be ultra-efficient, and the sheer speed of the trains allowed people to move great distances in a fraction of the time that it would have taken one hundred years prior.

The train passing areas scattered throughout the major train lines were a simple concept: adding a second vertical level of track that a train could access while another train whizzed by on the main track.

The train switching capability was modeled after the archaic steel railroad cars and upgraded from horizontal to vertical switching. A train flagged to yield by traffic controllers would simply slow down for the passing area. It would be switched vertically to the second track level via the electromagnetic rails. After the lower train had passed in the other direction, the first train would accelerate and decline again to the main track via the same magnets.

In busy city centers, there were several different vertical levels of tracks, including above and below-ground switching tracks. The trains for transport between active city centers spanned hundreds of meters in length but could be collapsed to be much smaller for lower population areas. The elimination of fossil fuels and the discovery of fusion technology meant that America's transportation relied solely on clean, renewable energy.

A high-speed train or ELT (Electromagnetic Levitation Train) from New York to Los Angeles took only eight hours and could transport eighty people with luggage per train car, and the ELT could pull thirty-two cars in one trip for a total of 2,560 people per train. One ELT moved the equivalent of more than five high-capacity aircraft making the same trip. When the population travel patterns throughout the country had become predictable with constant AI monitoring of the flow of people on any given day, the ELTs moved a million people per day.

The high capacity of these ELTs meant that people could travel for a fraction of the cost of traveling on an airline and not damage the environment in the process.

Freight transportation throughout America had become ultra-efficient as AI became more and more adept at predicting and optimizing freight movement patterns. Very long convoys of electric transport trucks filled with food and parcels traveled the freight transportation lanes on the roadways every day of the year. These designated freight lanes saw virtually endless, modern, black container trucks traveling between city and rural centers twenty-four hours a day, seven days a week.

The truck convoys traveled in long, aerodynamic lines that allowed for increased efficiency due to drafting. Drafting was used, such that the first truck in the convoy was pushing the maximum amount of air as it traveled. Still, the subsequent trucks in the convoy, who were within one meter of the truck in front, had to push significantly less air so that most of the convoy was continually in the truck's slipstream in front of it. Even the color of the trucks was chosen to give the maximum energy efficiency for travel; all the trucks were equipped with solar panels that charged the batteries using sunlight, but the black color collected radiant heat from the sun.

Radiant heat was collected more by the black color, as black absorbed all the different wavelengths of colored light and created ambient heat. This ambient heat was then used to heat the compartment in winter conditions and to convert the heat to battery energy. The process of transforming ambient heat into battery power used tens of thousands of capacitor-like devices that used the expansion property of specially designed materials. Once the capacitor was full, it released the harnessed energy to the power banks of the train.

Road maintenance for electric passenger cars and freight transports was completed entirely by robots; data on road conditions such as potholes or washouts was constantly fed through the lead freight cars to a central processing area where priority was assigned based on the importance of the route, and the severity of repairs needed.

Large autonomous resurfacing machines as wide as a single lane were dispatched to the problem area, and they would scrape up the top layer of the existing road, heat it, mix it, and drop a newly paved surface out the back of the resurfacing unit. The technology for resurfacing was over one hundred years old; however, machines had taken the reigns and perfected it to a startling degree. Depending on the existing material on the roadway, varying viscosities of heated, recycled plastics were added to allow for a solid and weather-resistant mixture.

By the early 21st century, humans had accumulated so much plastic waste that finding a place to keep it was a major environmental problem. Using existing waste plastics in the road surface compound allowed for a solid and long-lasting smooth surface on which to travel, and entire companies were founded to collect waste plastics from cities, landfills, and oceans to maintain roadways.

The scavenging of plastics had become a virtual gold mine for people who had the means to find, collect, and transport plastics to the Roadway Management System (RMS) depots, which had several major collection areas throughout North America. Once collected and sorted by varying qualities and compositions, the newly coined meta-plastics were distributed throughout the continent for road resurfacing. An entire army of robots working as one unit, from data given to them from a considerable distance, could function unassisted by humans almost indefinitely.

Robots maintained themselves; problems seldom arose that the few human supervisors at major materials plants or train traffic control units had to figure out for themselves.

After the majority of the world's plastics had been

collected, cleaned up, and repurposed, all newly produced materials were biodegradable. All biodegradable wastes were collected and diverted to extensive biodegrading facilities, which used organisms to turn the waste into a hot paste that boiled and bubbled in a compression chamber for days while producing large amounts of heat along with some methane gases.

Side units attached to the biodegrading units used the heat production and pressurization of the compartment to power generators that put large amounts of electricity onto the grid. The collected methane gas was compressed and stored in tanks which were distributed to cold areas of the continent. The gas was used for reheating in the road resurfacing process when extra power was needed due to cold ambient temperatures.

Humans viewed hourly, monthly, and yearly data on freight transport, road maintenance, and waste management as an oversight. Humans had finally taken the wheel when it came to preserving the planet, and it paid off in a big way in the late 2060s with the perfection of large-scale fusion energy. Power plants distributing solar, wind, hydroelectric, and geothermal energy worldwide were rapidly overshadowed by the massive amount of clean energy created with fusion technology.

Humans had used nuclear fission for almost a century before large-scale fusion energy was perfected. Fusion meant that there was no more radioactive waste, like in fission reactions, and fusion energy released nearly four million times more energy per equal mass than fossil fuel and four times as much energy as nuclear energy.

In the 2030s, there was a scientific breakthrough in the study of chemistry, particularly the discovery of a new isotope of hydrogen that previously seemed impossible. Like many scientific discoveries, Quadritium had been discovered by accident; before the discovery, Hydrogen-4 had been deemed a highly unstable isotope of hydrogen and was written off as a possibility to use in any fusion process.

By the 2050s, Quadritium was created in a laboratory by bombarding tritium molecules with fast-moving deuterium molecules; however, it was later discovered that by combining tritium molecules with deuterium in an oscillating magnetic field, the resultant molecules were stable enough to use in small-scale fusion reactors. In the following decade, the fusion reactors were scaled up to the high-yield ones used in America in the 2060s.

Humanity marveled at its accomplishments and started to reap the benefits of drastically reducing the pollution of their planet. Unfortunately, SG-2131 was discovered seven decades later and threatened to un-do humankind's advancements and accomplishments, changing humanity forever.

Chapter 59: Honoring a Promise

Mitch walked into the secure area of the HEPA lab Sunday morning at 6:50, as per his regular routine. He first went to the kitchenette and glanced at the workspace he and Fran shared as he walked past.

"Morning Fran!" Mitch exclaimed as he entered the kitchenette and fumbled through the cupboards for the desperately needed coffee. He had seen the back of Fran's head at her computer while walking past their work area.

"Morning, Mitch, good weekend?" Fran replied.

"Oh, you know, uneventful, really," Mitch lied, "Andrea and the kids are gone to her mom's for a few days; they'll be back on Tuesday." Mitch got the coffee grounds poured into the hopper of the coffee machine; he ground extra coffee grounds because he would need it today. The research area had unimaginable technology at its fingertips. Still, coffee was a do-it-yourself because they wanted to avoid another robot in the way at the HEPA lab to serve food and coffee.

"You want coffee or espresso?" Mitch yelled.

"Espresso, double, please," Fran yelled back.

"You got it," Mitch said as he put the cup under the machine and selected two espresso shots. The machine whirred as Mitch returned to the cupboard and found his mug. He looked at the mug.

World's greatest scientist, read the mug.

I don't feel like the world's greatest scientist today. Mitch thought as he put his mug under the second port of the coffee machine. A couple of button presses later, the machine whirred into life a second time, and Mitch grabbed Fran's espresso and walked back into their work area.

"An espresso for you," Mitch said as he put the cup on the side of Fran's workstation.

Fran didn't answer immediately; Mitch heard a distracted "Thanks" when he got halfway back to the kitchenette for his coffee. He grabbed his steaming cup of coffee from the coffee machine and walked back into their shared workspace.

"How about you, Fran? Good weekend?" Mitch sat at his computer and entered his password by selecting holographic words projected just above his desk. He looked at the camera perched above his workspace, which scanned his retinas.

Mitch leaned to the side of his workspace to see Fran's face. He recoiled as her face came into view.

"Eeek. Rough night?" Fran's eyes were bloodshot and sunken, and she had deep black bags under her eyes. She was visibly distressed. "Fran?" Mitch said as he wheeled his chair around to be on the same side of their adjoined desks as her.

Fran's unfocussed eyes left her desk and slowly landed on Mitch. She gave her head a shake as her eyes focused. She had been crying.

"Uh, yeah, weekend was okay. I went back to my godfather's place back home in Wigtown; we had a small ceremony and cleared out his house. It's going up for sale if you know anyone in the market in Wigtown," Fran lied. *It's more like I went back home to Wigtown to clear out a house I had never been to and whose owner I'd never met and talked to people pretending to be my relatives about my dead godfather, who doesn't exist,* she thought.

Mitch rolled closer to Fran, "Fran, I'm really sorry. I know this must be hard for you, losing your godfather and all. And taking care of his estate has got to be a lot on your mind."

"Yes," Fran started to play along, "Yes, it is a lot. But, after you get through your morning routine, let's get on that theta radiation experiment we worked on a couple of weeks ago. I have some ideas to run by you to get the condensing liquid to behave like we want it to."

"Yes, absolutely, Fran." Mitch was reminded about his tasks to do that morning.

He glanced at his watch. It was 7:03 a.m. Not that he needed any reminder of what he was supposed to do that morning. *No pressure, Mitch,* he thought as he stood up, wheeled his chair back to his desk, and started checking e-mails. Fran began writing on her virtual screen, but it was slow and deliberate.

Mitch took a deep breath and went over his instructions one last time.

He rolled his chair back from his desk. "Uh, I need

to switch up my portable drive unit." He held up his portable storage unit that hung from his neck on a lanyard. The unit was not only a mobile storage unit, but it also functioned as a security unit that tracked Mitch's movements while in the HEPA lab, allowing him to use all of the equipment and access all of the rooms without taking up time with retinal scans at each secure area within the lab.

As per HERB mission protocol, the PDUs needed to be recycled every two weeks or sooner if the wearer was conducting experiments and was using the PDU to log observations. Mitch conveniently had a reason to switch up his PDU before continuing his work with Fran on the theta radiation condensing liquid; the experimentation process generated vast amounts of data that were collected and funneled wirelessly via the PDU into large digital storage banks, where the HEPA AI unit could analyze the data.

Mitch walked past the kitchenette and down the empty hallway to the supply room. He was trying to act natural as he reviewed Leonardo's post-vomit instructions. He started humming a G-mob song, which sounded ridiculous, but it was the first song that came to mind, and it was on par with Mitch's brand of self-amusement while in the HEPA lab—kid's songs.

I'm so glad Andrea and the kids are out of town, Mitch thought as he approached the supply room. *I will be a complete wreck by the end of today with all the sneaking.*

Mitch got to the supply room and stopped directly in front of the door, and as the PDU verified his identity, the steel door gave an audible 'click,' and Mitch pushed the door open and stepped inside. He paused briefly, expecting the lights to come on, but they did not. Mitch exhaled as he stepped closer to the end of the shelf in the small room and felt for the port in front of the robot unit that was standing there, inactive. The heavy metal door closed behind him with an authoritative click. He inserted the PDU port into the front port of the robot; it was not hard to find. It was precisely in the center of the robot's chest, and Mitch put the PDU port into the robot's receptacle and counted mentally.

One quantum particle, two quantum particles, three quantum particles. Mitch pulled the PDU from the robot's port, stepped back, and manually pressed the light switch, and the lights came on in the room. He found the container of blank PDUs, grabbed one, left the supply room, and headed back to his computer. Mitch could feel a bead of sweat slowly moving down the side of his face by his temple. He started to hum again as he attempted to distract himself from the critical importance of what he was doing.

"G-mob Mitch?" Fran said as Mitch got within earshot.

"Well, when you hear the song a thousand times a day like I do in my house, it kinda sticks," Mitch retorted as he walked back to his desk, sat down, and placed the new PDU on his desk. With a few taps with his fingers on the virtual screen, the PDU was ready to be switched for the new unit. Mitch removed the lanyard from the old device with a click and moved it to the new unit with a similar click, and the new PDU was functional.

Once Mitch saw on the virtual screen that his new PDU was adequately set up, he got back up. He took a few steps to a small receptacle on the wall labeled "used PDUs." He dropped the unit into the slot, where the maintenance robot group would collect it, upload it, format it, test it for functionality, and put it back into the storage room.

Mitch turned from the PDU receptacle, "Okay, Fran, got my new PDU, and I'm ready for the radiation lab. What ideas did you have about the condensing liquid? Oops! I better log that malfunctioning light sensor in the supply room first." Mitch got back to his computer and, with a wave of his hand, brought up the virtual monitor; after selecting a couple of options, he was in the maintenance folder, and from there, he left an audio clip, "Light sensor in the small supply room is malfunctioning."

With a final wave of his fingers, Mitch had submitted the work order to the robot maintenance team, who would complete the task that night when no one in the HEPA lab would interrupt.

The director was watching the video and AI analysis of the HEPA lab that morning, and as he sipped his morning coffee, he watched the row of cameras that showed Mitch Smith and Fran O'Shea going about the start of their work day. He glanced down at the AI live interpretation, which the director requested with a particular focus on stress, deception, and concealment. Not surprisingly, Mitch and Fran had high stress levels, as the AI cameras captured breathing rates, pupil dilation, body temperatures, and heart rate.

Fran was showing low calculated levels of deception, which could have been falsely low due to her uncharacteristically weepy behavior in the lab over the last few days.

She's cracking, the director thought as he changed his attention to Mitch Smith. High stress levels were expected, but Mitch's deception and concealment levels were slightly higher than in the previous time. AI spectral analysis detected low levels of Acetaldehyde and acetate in Mitch's breath.

"Drinking again, are we, Mitch?" the director smiled. "You probably don't want anyone to know you're a stress drinker, right Mitch? Too bad Tactol doesn't eliminate bi-products of alcohol." The director's sarcasm was palpable.

The director swiped the images of Fran and Mitch off his main screen and closed his eyes briefly as he thought. He had set AI to monitor Fran and Maryam constantly, and nothing outside of the ordinary had been seen or heard from either of them after they left the director's office days ago. Nothing except Fran's spike in stress and recent tendency to break down in tears.

He brought Maryam's profile, a large panel of his holographic projection screen, to the center. Stress levels were up as well, but nothing else. She was going about her semi-retired life's business. Since there had been no further mission meetings, she went places with her husband, volunteered at the food bank, and drove sick people to and from medical appointments.

Sickening, the director thought in disgust, as he swiped Maryam's information away from his main screen. "How can they go about their meaningless lives knowing what is coming?" he grumbled.

"In a few days, Fran will be a write-off, and I'll be able to force her into taking a stress leave with all the video of her crying in the lab." He smiled, "she's not as strong as I thought. I guess being an academic her whole life and never having to struggle is making little Franny crack under the pressure. As for Maryam..." The director pulled at his goatee as he thought for a minute as AI followed and recorded his train of thought.

He pulled Mitch Smith's profile to the center as Maryam's receded into the background. He sighed as he examined Mitch through the eyes of AI. "Fran's whipping post, hates confrontation, stress drinking, young family life." He stared through the information as he thought. His eyes came back into focus as he smirked.

"How about you come and sit down with me, Mitch? Maybe the stress of Gus' death and Leo's disappearance have rattled you enough that you and Fran need to take about six weeks away from the lab. Stress drinking can't be good for your young family, and we need you to be at your best for the mission. Besides, we are at a point in the mission where we are simply waiting to see if we have been successful," the director mimicked, talking to Mitch.

He swiped his hand again as Mitch's profile disappeared, and the moon colony appeared: several large domes accommodated multiple communities, with greenhouses, small livestock, and community centers. He manipulated the view with his fingers and zoomed in on one of the more giant domes, then again to zoom in on a substantial residence at the dome's edge. He leaned back in his chair as he glanced at the population stats for the moon colony.

Population: 26,081: 13,871 female, 12,210 male.

He used his fingers to zoom in on the residence at the dome's edge. Over five hundred square meters of private space, and only about a kilometer from where humanity had spent the last few years accumulating DNA samples, animal and human zygotes, plant seeds, fusion generators, large rolls of electrical wiring, atmospheric scrubbers, robots of every kind, landing crafts, and computer drives filled with every bit of knowledge that humanity could think of to help should the mission to divert the asteroid fail.

The director had seen to it that he was the recipient of the lovely large house he was looking at, a gift from a wealthy benefactor. He was not taking any chances: he would be set for life, either on Earth or the Moon.

B.E. Smith

Chapter 60: The Moon, The Mine, and The Mother

The moon has intrigued and inspired humankind since people first looked up in the sky and saw our closest neighbor going around and around with clockwork regularity. In 1969, humanity landed a craft on the moon, drove on a moon rover, hit some golf balls, bounced around, and visited briefly five more times in the next three years. After 1972, no humans walked on the moon for ninety-one years until 2063, when a lunar colony had been started as a joint venture involving twenty-eight countries.

At first, temporary crews stayed for six months as they constructed the first permanent human residence near a sizeable frozen water deposit at the North Pole. Perpetua, the first permanent human colony on the moon, was named to represent humanity's aspiration and determination to perpetuate as a species, to survive and thrive as a species, even under challenging circumstances. Although it was a statistical guarantee at some time in the future, a significant asteroid strike was not the only reason to colonize the moon. The moon was seen as a mustering point for exploration of Mars and other planetary systems in the future.

In the event of a large asteroid strike, if humankind lost the use of their planet for hundreds of years or more, they wanted to come together as best they could to make sure that they survived long enough to re-inhabit the Earth. DNA sampling and sequencing from the twenty-eight countries that founded the colony had determined that 99.4% of human genetic material could be represented on the moon in the form of living people, DNA samples, or frozen zygotes.

In the months and years after the 2131 discovery of SG-2131, humanity systematically started choosing people from all walks of life and all cultures, academics, philosophers, engineers, farmers, geneticists, leaders, and workers to populate and expand the moon colony.

This was to become Earth's proverbial Noah's Ark, where any and every organic life form's DNA was captured and stored for re-population of the Earth. Scientists in several countries created Earth Arks (EA) to keep their histories and cultures. One such country was Canada, which not only had a vast land mass but also some of the most stable land in the world.

The Canadian Shield was a source of mining materials for centuries, and humans became proficient at creating underground pathways and storage areas. One such storage location in a Northern Ontario community was founded by mining explorers, then converted to tourism, and then used for scientific research into ancient water samples.

The next stage for the mine shaft system was unexpected; however, the Timmins, Ontario community embraced the challenge to create the only genetic and information assembly unit (GIA) in the country and only one of two in North America. GIA was over two kilometers below the Earth's surface and boasted a waterproof, heat-resistant, and shockproof outer layer that would keep the precious substance of humanity safe until it could be recovered by distant moon-dwelling ancestors.

The ingenious shockproof layer surrounding the entire facility was five meters thick and housed dampened gel-filled spaces every one-half meter, designed to break up shock waves that ripped through the Earth's crust.

Designed in a large oblong sphere, like an egg, you could see the layers of rigid wall side by side with the multiple layers of dampening gel by slicing into the egg. By breaking up the shock waves with gel, the propagating waves would be minimal or non-existent by the time they got to the actual outer shell of the GIA.

Inside the layers of shock, heat, and cold resistance, the GIA was suspended on a gigantic series of dampening coils. High temperature-resistant drainage throughout the GIA layers ensured that if large amounts of water or lava penetrated any layer, it would be directed away from the inner storage. In short, earthquakes, lava, or water could assault the GIA, and most, if not all, pods would survive to be re-discovered. The only way that the Canadian GIA would not survive was if the asteroid touched down close enough to crush or eject the pods.

The inside of the GIA was something out of a sci-fi movie: long rows of independent compartments called pods that were waterproof, heat resistant, and shock resistant themselves, each labeled with the name of its prized contents. There were two GIA Prime pods, one at each end of the underground lair, each a copy of the other. They provided the interface with which humans could access all of the pods in the future. They controlled the robots left to do all of the organization and heavy lifting as humanity got ready to seal its time capsule.

Upon completion, all of the layers would be sealed by robots starting at the outside layer, moving inwards, and the internal clock of the GIA would start counting its slumber. Atmospheric instruments placed at ground level fed weather conditions to the GIA and stored them in its memory. However, it was widely recognized that the initial touchdown of SG-2131 would most likely obliterate any sensory equipment on the surface. To this end, GIA had a small group of robots whose sole purpose was to tunnel to the surface every fifty years and transmit atmospheric and ground conditions to the Moon-Earth observatory.

As each subsequent robot surfaced from the GIA, the following generation of humans received the message and established contact with the robots. The robots could upload new data or instructions, such as more recent software or a change in transmission frequency, as the Earth became more and more habitable. The sweet spot, as it was called, was when humans could survive on the Earth unaided; however, by the time the sweet spot arrived, humans would have already established new colonies at the equator.

The generally accepted theory then was that the asteroid event would eventually plunge the Earth into an ice age due to the kilometers-thick layer of ash and water vapor blocking the sun. This was, of course, after the surface of the Earth was roasted by tens of thousands of cubic kilometers of lava, evaporating much of the oceans and lakes into steam into the atmosphere. This sauna-like effect would make the surface of the Earth inhabitable with superheated fog, followed by freezing over the next decades. So, the Earth would be flash-roasted and then slowly frozen.

Depending on where the strike occurred, lava could continue to pour out of the Earth's wound for decades. Except for some hardy micro-organisms, the events were expected to kill every living species on Earth. The theory of creating the first human re-colonies at the equator was that the equator would provide the earliest temperate surface conditions, and humans would not be subject to drastically swaying cold nighttime temperatures that would still plague the North and South of the planet.

Humans would have to re-create their moon dome communities at first, as the air quality and air composition would most likely be unsuitable without air scrubbers and dome structures. However, the scientific community and the myriads of computer simulations could not agree on how bad the air quality would be at that time.

In the best-case scenario, the frigid cold of the Freezing Age would allow most of the microscopic debris in the atmosphere to be dragged down to the ground by the changing density of the water molecules. An epic frost would cleanse the atmosphere of waste, and if not all oxygen-producing plants and algae had been wiped out, then the planet's oxygen levels would slowly return to normal.

Humans had witnessed the miraculous ability of nature to survive, from the hottest volcanic eruption to the coldest places on the Earth, and SG-2131 would test Mother Nature to her limits.

As with anything in experimental physics, the final result would depend on the exact initial conditions, which would only be revealed once the asteroid hit the Earth. Humans were well positioned to see the destruction of their home planet from the moon, and as the surviving human life waited and watched, for the first time in history, they lived and wrote a significant part of their history solely from the moon.

That is if the HERB mission was unsuccessful.

Chapter 61: The Intruder

In the early mornings, when Kevin was sleeping, Leonardo had gotten into the habit of walking a loop around the tiny volcano-top community. He had become much more acclimatized to the high altitude in recent days. Leo's drive to exercise was rooted solely in his need to become more involved in the mission, and to do that, he needed to be in better shape. Not only did he need to become more engaged in Kevin's research, but he also needed to be constantly on the lookout for Shadow.

He was beginning to enjoy walking again as his body acclimatized to the altitude, and his facial and head injuries became less intrusive in his daily life. Walking had always allowed Leonardo to clear his head and think.

He was absorbed in one of AI's generated timelines to insert himself into Kevin's research when he rounded the corner of the dormitory building and heard an unusual noise. Leonardo's head turned, followed by his body as he squared himself to the noise source. He froze as his senses frantically searched where the noise had come from. The morning twilight revealed nothing.

There was nothing but quiet. Then, Leonardo's blood pressure shot up as he felt his watch vibrating on his wrist. He glanced at his wrist as he felt the panic rising inside him.

Perimeter breach, target 50 meters East.

Leonardo glanced around him into the dimly lit shadows. There were no people outside, nor were there any lights in the dormitory windows facing him. He accepted the risk and raised his watch to his mouth as he whispered, "Identify target."

Unknown person. No active nanotech signatures.

Leo's eyes strained to focus in the direction of the unknown person as he slowly raised his hand to his mouth again. "Ready NDI, and wait for my command," he whispered.

Copy, read his watch.

Leonardo started walking slowly towards the darkly clothed person who was either sitting on the ground or was hunched over; he couldn't tell. Multiple scenarios were racing through Leo's head as he slowly closed the distance from the person, step by step.

It's a trap. I know it. Leo thought.

"Scan the rest of the area for a secondary breach," he whispered as he squinted at the black blob. His mind was trying hard to make out a shape.

No secondary breach. AI responded as Leo glanced at his watch.

Should I run and warn Kevin? Should I go closer? Could this be a distraction technique? Leo's thoughts were galloping as he continued toward the unknown person.

Could it be a person that needs help? That's what could make this a trap.

The suspense was killing Leonardo as his voice broke the silence. The tension in his voice scared him.

"Hey!" he yelled. "Identify yourself!"

There was sudden movement as the person dressed in dark clothes got onto their knees and started crawling, then got up and tried to run before falling awkwardly.

"AI, can you identify that person now?" Leo yelled, not even bothering to whisper this time. He looked at his watch impatiently, even though the response was almost instantaneous.

Possible: Bruce Hayden was AI's written response.

"Shit, shit, shit," Leo muttered to himself. He wanted to run and help if the person was Bruce, but his mind was still trying to figure out if it was a trap.

Bruce's height and build are similar to Shadow's.

"AI, stay ready to fire NDI," Leonardo yelled as he ran toward the person lying on the ground. Leo could feel his lungs struggling to keep up as he ran, but the adrenaline pulsating through his body urged him to keep going. He skidded to a stop and landed on his knees as he reached the person lying face down on the ground; the person was breathing, but the breathing was ragged.

Leo yelled, "Who are you?" as he shook the person and gripped the dark-colored sweater. He could see the short dark hair of the person as he pulled their hood back, revealing the back of the person's head.

Finally, the suspense was too much after the person did not react, and Leo grabbed two handfuls of the sweater and flipped the person onto their back. He gulped as his mind raced to keep up with what he saw. He felt like he was watching himself looking at the unconscious person. The person had a large, red, irritated-looking mark on his cheek, but otherwise, it was unmistakably the face of Bruce Hayden.

"Bruce, Bruce!" Leonardo yelled as he shook the man.

"AI, scan!" he yelled, "is this Bruce Hayden?"

The short vibration on Leonardo's wrist revealed the man's identity.

Confirmed. Bruce Hayden.

Leo started breathing harder as he looked Bruce up and down for an injury.

"Bruce, Bruce!" Leo yelled as he searched Bruce for any apparent injury. There was no response.

"AI! Scan for vitals and for injuries!" he yelled as he put his head on Bruce's chest, waiting to feel his chest rising. His watch vibrated, but Leonardo was so focused on waiting for Bruce to breathe that he didn't feel the vibration.

He wasn't breathing.

"No, no, no, no, no!" Leonardo was mumbling as he took his head off of Bruce's chest and put his index and middle fingers on the side of Bruce's neck to check for a pulse in his carotid artery.

Leonardo kept his fingers there as he closed his eyes, trying to distinguish his racing blood and Bruce's pulse. He remembered that he had not heard back from AI about his vitals. Leonardo sat up and checked his watch. He cursed himself for leaving the tiny earpiece to hear AI speak on his nightstand. He braced himself for what came next.

Not breathing, pulse rapid, and weak—146 beats/minute.

The text disappeared, and a new message was displayed.

Clear airway.

"Shit, shit, shit, shit, shit." Leo panicked as he straightened Bruce's head and leaned it backward to open his airway, and he opened Bruce's mouth, plugged his nose, put his mouth on Bruce's, and gave a small puff of air.

No obstruction, he thought as he gave a quick second puff of air.

He started to look down at this watch again for instructions from AI, and he noticed that Bruce's lips had turned distinctly blue.

"No, no, no, no, no, no, no," Leo yelled as he leaned down and gave Bruce another puff of breath. This time, he watched Bruce's chest rise slightly with the air input.

Still no response.

As Leo was frantically trying to read his watch, he could see the sides of his vision closing in on him; he shook his head as if to dissuade the darkness from continuing.

"Help! Help!" Leonardo screamed a broken scream as he tried to focus on Bruce. His eyesight was diminishing by the second. A calmness overcame Leonardo as consciousness began to withdraw from his body; he felt like he was floating over Bruce, just observing, not worrying or trying to help.

The sound of a massive gasp of air brought Leonardo's mind back into his own body.

Is it my breath? Is it Bruce's breath?

Leonardo's sluggish eyes tried to focus on the man lying down on the ground below him. He was trying to remember who the person was; he knew the man well but couldn't think of his name or why he was on the ground.

There was another gasp. This time, the man on the ground opened his eyes with a gasp. He sat bolt upright, and as Leonardo fell sideways, the man spoke. Leo's body didn't react to the sudden loss of balance, and he watched quizzically as his view turned sideways, and he landed on the ground with a thud.

"Leo!" the man croaked, his voice was coarse and raspy. He rolled Leonardo onto his back and waved his hands before Leo's eyes.

"Leo! Stay with me, man!" he croaked.

As Leonardo's mind slowly captured more and more of his surroundings, he realized who the man was.

"Bruce! Am I glad to see you!" he stammered.

"Leo! Are you ok? You're not going to believe who I ran into!" Bruce yelled as he patted Leonardo on his chest to reassure himself that it was Leonardo in the flesh. Bruce was pulsing with adrenaline, his chest heaving as he tried to calm his breath.

"Shadow?" Leo whispered.

"Yes, Leo, Shadow, I ran into Shadow," Bruce said as he sat Leonardo upright. His balance was still off, but he was coming to.

"Who is he?" Leo whispered.

Bruce looked at Leo as Leo waited for a reply. "Shadow is not a he, Leo. Shadow is a she."

"What??" Leonardo wheezed, "a woman?" he asked.

"Yes, Leo, Shadow is a she. I ran into her when I was trying to get to LAX. I barely escaped with my life. She's had face-altering surgery, but I'll need to access your AI unit to verify with certainty who she is. I hope my AI unit got good observations of her before it was disabled."

Leo looked at Bruce, trying to understand what Bruce was telling him. He was confused.

"Leo, Shadow is Lee Cain. You know, Leslie Cain?"

"What?" Leo gasped. "You mean, Lee, like Gus'

niece? I don't understand!" He held his hand up to the left side of his face. *It ached.*

"Yes, Leo. Lee Cain, Gus Maxwell's niece. The same Lee Cain, who I narrowly beat out to come on this last mission?" Bruce grabbed Leonardo's face as the two men spoke like he was trying to help the information sink in.

"But," Leonardo thought for a moment. "But why would she kill her own uncle? Why would she kill Gus? I don't understand?" He shifted his weight as he sat and looked past Bruce with a panicked look.

"No, Leo, she's not here. Just take deep breaths. Please, let's just get inside where we can get warm and get some food and water, and we'll figure all of this out."

Leonardo reached out and touched the red splotch on Bruce's right cheek.

"Yes, Leo, it was AI's idea to modify my face."

Leonardo smiled as he looked at Bruce. He just stared and smiled a big, broad smile.

"Ok, goofball, I'm happy to see you too." Bruce reached out and pulled Leo close for a long hug. They stood up arm in arm to support each other and started walking to the dormitory.

"I'm so glad you're here, Bruce," Leo whispered; his voice still hadn't returned to him properly.

"Me too, Leo. You have no idea. wait until you hear what I've gone through to get here." Bruce paused, "Kevin is good?"

Leonardo nodded awkwardly as they got close to the entrance to the dormitory apartment.

Bruce patted Leonardo's back, "Leo, let's go discover that asteroid."

To be Continued...